Festival and Event Tourism

Building Resilience and Promoting Sustainability

Festival and Event Tourism

Building Resilience and Promoting Sustainability

Edited by

Anukrati Sharma
University of Kota, Rajasthan, India

Jeetesh Kumar
Taylor's University, Subang Jaya, Selangor, Malaysia

Bakhodir Turaev
Tashkent State Transport University, Uzbekistan

Priyakrushna Mohanty
Garden City University, Bangalore, India

CABI

CABI is a trading name of CAB International

CABI
Nosworthy Way
Wallingford
Oxfordshire OX10 8DE
UK

Tel: +44 (0)1491 832111
E-mail: info@cabi.org
Website: www.cabi.org

CABI
WeWork
One Lincoln St
24th Floor
Boston, MA 02111
USA

Tel: +1 (617)682-9015
E-mail: cabi-nao@cabi.org

A catalogue record for this book is available from the British Library, London, UK.

References to Internet websites (URLs) were accurate at the time of writing.

ISBN-13: 9781789248661 (hardback)
9781789248678 (ePDF)
9781789248685 (ePub)

DOI: 10.1079/9781789248685.0000

Commissioning Editor: Claire Parfitt
Editorial Assistant: Lauren Davies
Production Editor: Shankari Wilford

Typeset by Exeter Premedia Services Pvt Ltd, Chennai, India
Printed and bound in the USA by Integrated Books International, Dulles, Virginia

Contents

Contributors

Akay, Bayram, Department of Tourism Management, Burdur Mehmet Akif Ersoy University, Burdur, Turkey. Email: bakay@mehmetakif.edu.tr

Apleni, Lwazi, University of Zululand, Department of Recreation and Tourism, KwaDlangezwa, South Africa. Email: ApleniL@unizulu.ac.za

Arora, Manpreet, School of Commerce and Management Studies, Central University of Himachal Pradesh, Dharamshala, India. Email: arora.manpreet3@hpcu.ac.in

Balasubramanian, Kandappan, Centre for Research in Tourism (CRiT), School of Hospitality, Tourism & Events, Taylor's University, Subang Jaya, Selangor, Malaysia. Email: Kandappan.Balasubramanian@taylors.edu.my

Bristow, Robert, Geography, Planning and Sustainability, Westfield State University, Westfield, MA 01086, USA. Email: rbristow@westfield.ma.edu

Chowdhary, Nimit, Department of Tourism and Hospitality Management, Jamia Millia Islamia, New Delhi, India. Email: nimitchowdhary@gmail.com

Coomansingh, Johnny, Independent Researcher/Cultural Geographer, Newark, Delaware, USA. Email: nanaba7@gmail.com

Figueira, Victor, Polytechnic Institute of Beja (IPB), Beja, Portugal. Email: victorfigueira@ipbeja.pt

Garcia, Ana Rita, Polytechnic Institute of Castelo Branco, Castelo Branco, Portugal. Email: anarita@ipcb.pt

Güven, Yiğit, Institute of Social Sciences, Ankara Hacı Bayram Veli University, Ankara, Turkey. Email: ygtguven@gmail.com

Hasana, Uswathul, Department of Tourism Studies, School of Management, Pondicherry University, Puducherry, India. Email: uswathulhasana2512@gmail.com

Henama, Unathi Sonwabile, Tshwane University of Technology, Department of Tourism Management, Pretoria, South Africa. Email: HenamaUS@tut.ac.za

Hussain, Kashif, School of Media and Communication, Taylor's University, Subang Jaya, Selangor, Malaysia. Email: Kashif.husssain@taylors.edu.my

Jenkins, Ian, Faculty of Life and Environmental Sciences, Háskóli Íslands/University of Iceland, Reykjavík, Iceland. Email: jenkins@hi.is

Konar, Rupam, Centre for Research in Tourism (CRiT), School of Hospitality, Tourism & Events, Taylor's University, Subang Jaya, Selangor, Malaysia. Email: Rupam.konar@taylors.edu.my

Kumar, Jeetesh, Centre for Research in Tourism (CRiT), School of Hospitality, Tourism & Events, Taylor's University, Subang Jaya, Selangor, Malaysia. Email: Jeetesh.Kumar@taylors.edu.my

Kurnaz, Hande Akyurt, Department of Tourism Guidance, Faculty of Tourism, Bolu Abant İzzet Baysal University, Bolu, Turkey. Email: handekurnaz@ibu.edu.tr

Lavandoski, Joice, Universidade Federal do Estado do Rio de Janeiro (UNIRIO), Rio de Janeiro, Brazil. Email: joice.lavandoski@unirio.br

Machado, Annaelise Fritz, Polytechnic Institute of Cávado and Ave (IPCA), Barcelos, Portugal. Email: annaelisefritz@yahoo.com.br

Makuyana, Tawanda, Tourism Research in Economics, Environs and Society (TREES), North West University, Potchefstroom, South Africa. Email: makuyanatawanda@gmail.com

Mohanty, Priyakrushna, Garden City University, Bangalore, India. Email: pkmohanty90@gmail.com

Mothiravally, Vinayaraj, Dimension International College, Singapore. Email: mothiravally.v@dimensions.edu.sg

Nunes, Sara Morgado, Polytechnic Institute of Castelo Branco, Castelo Branco, Portugal. Email: sara@ipcb.pt

Nzo, Zukiswa Yoliswa Zinhle, National Council of and for Persons with Disabilities, Edenvale, South Africa. Email: nzo.zukiswa@gmail.com

Ön, Funda, Department of Tourism Management, Faculty of Tourism, Mugla Sıtkı Kocman University, Mugla, Turkey. Email: fundaon@gmail.com

Phale, Jankie Kgalabi, Tshwane University of Technology, Department of Tourism Management, Pretoria, South Africa. Email: PhaleJK@tut.ac.za

Sharma, Anukrati, Department of Commerce and Management, University of Kota, Rajasthan, India. Email: dr.anukratisharma@gmail.com

Sousa, Bruno, Polytechnic Institute of Cávado and Ave (IPCA), Barcelos, Portugal. Email: bsousa@ipca.pt

Swain, Sampada Kumar, Department of Tourism Studies, School of Management, Pondicherry University, Puducherry, India. Email: sampadswain@gmail.com

Tiwari, Pinaz, Department of Tourism and Hospitality Management, Jamia Millia Islamia, New Delhi, India. Email: tiwaripinaz@yahoo.com

Toksöz, Derya, Isparta University of Applied Sciences, Isparta, Turkey. Email: derya.toksz@gmail.com

Turaev, Bakhodir, Tashkent State Transport University, Uzbekistan. Email: profturaev@gmail.com

Vareiro, Laurentina, Polytechnic Institute of Cávado and Ave (IPCA), Barcelos, Portugal. Email: lvareiro@ipca.pt

Yayla, Özgür, Department of Recreation Management, University of Akdeniz, Antalya, Turkey. Email: ozguryayla@akdeniz.edu.tr

Yozukmaz, Nisan, Department of Tourism Management, Faculty of Tourism, Pamukkale University, Denizli, Turkey. Email: nisany@pau.edu.tr

Introduction

Anukrati Sharma[1]*, Jeetesh Kumar[2], Priyakrushna Mohanty[3] and Bakhodir Turaev[4]

[1]University of Kota, Rajasthan, India; [2]Taylor's University, Subang Jaya, Selangor, Malaysia; [3]Garden City University, Bangalore, India; [4]Tashkent State Transport University, Uzbekistan

Events are an integral part of the tourism system. They constitute major elements both at the origin point (events as push factors to travel) and at the destination point (an event as a pull factor in marketing and development of the destination). At a destination level, event tourism develops and markets planned events as tourist attractions, catalysts, animators, image makers and place marketers. This process includes bidding on, facilitating and creating events and managing portfolios of events as destination assets. For individual events, event tourism means taking a marketing orientation to attract tourists, sometimes as an additional segment and sometimes as the core business. Destination events are organized to entertain the tourists as being the core business. From the demand side, event tourism refers to travel propensity to attend events. This includes both dedicated event tourists who are motivated to travel for specific events and other tourists who attend events while away from home. Festivals, exhibitions and special events conducted at a large scale serve the purpose of bringing economic benefits to the host economy and create a favourable positive image of the place(s) hosting them.

It is argued that due to their ability to attract a large number of tourists and their spending, event tourism has emerged as a strategic area for various destination managers and is being exploited as a tourism asset to boost tourism at a particular destination. Further, the sustainability of event tourism has become a rising concern given the narrow spatial and temporal nature of events resulting in challenges like negative image in the public sphere, lack of participation opportunities and distorted dialogues among community members. Literature supports that while festivals and events come up with ample benefits for the hosting community and destination, numerous challenges must be addressed to make these benefits and the destination sustainable.

This book is essential from many aspects, especially after the coronavirus disease 2019 (COVID-19) pandemic. As we all know, festivals and events are integral parts of our culture and lives. Currently, as the whole world faces an economic crisis, festivals and event tourism can play a critical role in putting the tourism industry back on the path to recovery. This book will focus on how festivals and event tourism can support tourism and community development. Festivals and events are more based on relationships than sales, and they are more enhanced through talent and skills. While it is not necessarily easy to enter the events tourism market, it provides ample prospects to various stakeholders to earn a livelihood.

*Corresponding author: dr.anukratisharma@gmail.com

To speed up the recovery and redesign of the tourism industry, we need to push festival and event tourism across the globe. The book does not deal exclusively with the COVID-19 pandemic, but its focus is on the recovery options/opportunities from the said pandemic. Most countries are more or less facing similar problems, and one of the biggest challenges is the economic problems. The local community, directly or indirectly involved in the tourism and hotel industry, will suffer massive negative impacts from the economic crisis. Restarting the shut hotels, travel agencies and restaurants will need many investments that do not seem feasible after the pandemic. That is why the editors picked festivals and events as a resource to mitigate the after effects of the pandemic in an economic, social and psychological way.

Recently, many research studies have been conducted outlining the importance of building resilience for events, tourism and festivals during and post-COVID-19. However, promoting sustainability for events, tourism and festivals in a more comprehensive manner has attracted very few researchers. Also, not much literature is available on possibilities and challenges associated with festivals and events in achieving destination sustainability. Considering the ongoing trend of destination sustainability in events tourism, few research attempts have explored diverse aspects.

This book is divided into four parts. Part 1 of the book includes the aspects of sustainability in events, issues and concerns raised due to pandemic and post-pandemic management of the event industry. Part 2 discusses the concept of green events and festivals and green service innovation in the event industry. In Part 3, the impact of festivals and events on local communities is the focus. Part 4 includes case studies on event and festival tourism destinations and events as tools for destination branding and promotion. With its 15 chapters, this book contributes to the knowledge in this identified research area. A brief outline of the contents of each chapter is as follows.

Chapter 1 by Toksöz includes an overview of recent issues affecting the resilience of the events and assesses their role in the sustainability of the events. The chapter also includes ways to overcome negative situations and bring more resilience to the events industry.

The authors of Chapter 2, Mohanty, Sharma, Kumar and Turaev, introduce a framework for a resilient and sustainable tourism recovery plan post-COVID-19. The chapter concludes that a good recovery plan will continually reinforce the positive strengths and minimize the vulnerabilities to make the system sustainable.

In Chapter 3, the author Coomansingh focuses on the use of parang and its creation of the annual local festival *La Fiesta de Lopinot* for domestic tourists to enhance the economy of the village of Lopinot through tourism in Trinidad.

The authors of Chapter 4, Hasana and Swain, systematically review the literature to explore the opportunities and limitations of small-scale events in achieving economic, social and environmental sustainability. The chapter concludes that small-scale events can influence the growth of the tourism sector compared to mega-events.

In Chapter 5, the authors Ön, Kurnaz and Yozukmaz report on an analysis of the opinions of the tourist guides working in seven different geographical regions of Turkey to discuss cultural sustainability, which is a part of the notion of sustainability in tourism. From their findings they conclude that as a tool of culture, festivals are very significant for cultural sustainability but, at present, the full potential of their contribution to achieving cultural sustainability is not being met as those involved in festival organization do not see the importance of cultural sustainability.

Chapter 6 by Jenkins and Bristow reports on the findings from attendees of Iceland Airwaves music festival to ascertain the visitors' ranking of sustainability, and offers strategies for music festival management to become more resilient. The chapter concludes that Airwaves should follow the lead of the 2012 Olympics in London and assess their impact using the International Organization for Standardization (ISO) standards. ISO 20121 *Event sustainability management systems – Requirements with guidance for use* (ISO, 2012) is designed to help planners address sustainability for everything from a music festival to the Olympics.

In Chapter 7, Machado, Sousa, Lavandoski, Vareiro and Figueira describe how they used semi-structured interviews and content analysis with Rock in Rio participants to compile a list of best practices in managing music festivals with environmental concerns in a sustainable way.

The authors of Chapter 8, Konar, Hussain, Mothiravally and Balasubramanian, describe a comparative study they conducted on green innovation performance in small and medium-sized enterprises (SMEs) organizing events across Malaysia and Singapore. They examine the contextual effect of organizational absorptive capacity on green innovation performance.

The authors of Chapter 9, Makuyana and Nzo, present a framework for a philanthropic event that increases accessibility and opportunities for disabled people to participate in community development in South Africa. The chapter nurtures a narrative of enhancing communities' understanding and relationships with disabled people as part of an inclusive society.

In Chapter 10, the authors Nunes and Garcia discuss the results of two studies, assessing the perception that the inhabitants of Idanha-a-Nova have about the Boom Festival in 2009 and in 2021. The chapter confirms that the Boom Festival is perceived as positive for the local economy as it attracts many visitors and promotes the region internationally. However, less positive aspects were also mentioned, such as the consumption of drugs and alcohol, increased prices of goods and services, and traffic congestion.

In Chapter 11, Apleni, Phale and Henama report on an investigation of events promoted in rural areas of South Africa and the Kingdom of Lesotho that have benefitted the geographic spread of tourism. These community-initiated events have grown organically, attracting more visitors and aspects of commercialization, and have become hallmark events in their localities.

In Chapter 12, Akay examines the Kırkpınar Oil Wrestling Festival, identified by the United Nations Educational, Scientific and Cultural Organization (UNESCO) in 2010 as an Intangible Cultural Heritage of Humanity. The chapter focuses on the relationship between sports heritage events and tourism.

The author of Chapter 13, Arora, discusses the role of women in Indian festivals. Although some critics are of the view that festivals are actually celebrated to exhibit patriarchy, the author contends that the basic fundamental principle of celebrating life in Indian tradition is actually to give centrality to women. Through festivals, rituals and various religious ceremonies, it becomes evident that women are inseparable from festivals. The fundamental essence of these celebrations totally, purely and incredibly lies with the women, and their existence is the fundamental basis of these festivals.

In Chapter 14, the authors, Yayla and Güven, examine the concept of event tourism and destination branding. They also include the need to use technology to promote the brand of a destination at different levels, including that of the country, institutions and organizations.

In Chapter 15, the authors Tiwari and Chowdhary discuss the role of technology in managing crowds at events, in particular in the context of the Kumbh festival in India. Overcrowding at events is a cause for concern as it negatively impacts on the experience of tourists and travellers, resulting in negative evaluation of the event and the destination. Technology can be used to help with monitoring and managing the flow of visitors at events.

Reference

ISO (International Organization for Standardization) (2012) *Systèmes de management responsable appliqués à l'activité événementielle – Exigences et recommandations de mise en oeuvre* [*Event sustainability management systems – Requirements with guidance for use*]. ISO, Geneva, Switzerland.

Part 1: Transforming Sustainability into a Strategy

1 Making Events More Resilient: Emerging Issues and Concerns

Derya Toksöz*
Isparta University of Applied Sciences, Isparta, Turkey

Introduction

Events are held today in many parts of the world and in a wide variety of genres. The events, which attract many attendees on a national and international scale, need to be evaluated in terms of their sustainability. Although the foundations of the concept of sustainability were laid for the protection of natural elements, today's understanding and approach to sustainability has developed and it has been used in almost every aspect of social science and life (Hall, 2012). Until recently, little attention has been paid to the impact of planned events on the sustainability and resilience of destinations. Many events leave a strong social and cultural legacy in the destination where they are organized. Events often involve the creation of new infrastructure; they may be arranged in environmentally sensitive locations, require large volumes of participants, goods and services, and other scarce resources to be transported, causing disturbance to local residents (Jones, 2017).

When the culture of participation in events is examined it is seen that the increasing interest of society in mega-events such as the Olympics, the World Cup and the Super Bowl has started the development of the event literature (Backman, 2018).

After World War II, transglobal travel became easier and cost less and workers in new technologically developed countries had more discretionary income and increased leisure time and this caused an explosion of global events at all levels. An increase in the number and frequency of events of all kinds began to appear (festivals, sports, music, food, culture, etc.) (Getz, 2008). Events include many types of decision making before, during and after the event. Event organizers are beginning to realize that events involve a more systematic approach that includes strategic planning that includes not only short-term success but also long-term benefits for the wider community (Zrnic and Susnjar, 2010).

The broadest area of research focus has been the economic effects/benefits of individual events from the beginning of event research. The second group of event research topics is located around visitors attending various events. It has been found that research is weighted to understand what motivates the needs and desires of visitors to participate in the events (Backman, 2018, pp. 169–170). Today, there is increasing militarization and security, which aims to protect participants and of course tourists from local hostilities and security-related issues, especially terrorist attacks.

One of the important issues in the success of an event depends on the establishment of

*derya.toksz@gmail.com

DOI: 10.1079/9781789248685.0001

risk management in the context of strategic and operational problems. Another important issue is technological development and its impact on the event sector because technology is reshaping the sector as well as the demands and expectations of participants (Eventbrite, 2013). Prior to advances in event technology, event planning was a manual process. Similarly, due to the increase in availability of new technology and the coronavirus disease 2019 (COVID-19) pandemic today, it has made it easier to bring personal and virtual activity ideas to life (Bizzabo, 2021). Advances in data collection, analysis and communication technology greatly improve the way events are planned and managed.

In a short time, the world agenda has become very intense due to the developing technology, terrorist attacks and finally the COVID-19 pandemic. The world has experienced a number of major epidemics/pandemics over the past 40 years, but none have had similar consequences on the global economy and tourism as the COVID-19 pandemic (Gössling *et al.*, 2020). For example, the impact of cancelled events, closed accommodation and closed attractions was immediately felt in parts of the supply chain, such as food, laundry and entertainment centres (Business Insider, 2020). Event managers trying to find solutions for the extensive challenges in their marketing efforts, for example, increased media and audience fragmentation, and this resulted in greater competition in the media market (Martensen *et al.*, 2007).

This chapter aims to provide an overview of current issues, concerns about events, and what ways events can be used to overcome negative situations and be more resilient. With regards to events, it is important to evaluate what the practices and trends are about, both for the participants and for the organizers in this process. The main aim of the chapter is to examine the recent issues affecting the resilience of the events and assess their role in the sustainability of the events. For this purpose, the literature has been searched and recent studies on sustainable practices have been examined. The events industry is still developing the research relationship between events and the environment, and is working on how making events more resilient is relevant to the sustainability framework (Rojek, 2014).

Emerging Issues about Events

Most of today's tourist destinations use events and festivals, one of the most important trends in the pursuit of differentiation. These countries consider the events truly the best for their destination and best suited to the existing infrastructure, markets and products (Van Niekerk, 2017). The changes and concerns that have arisen in the planning and management of events over the past decade are a must for a responsible event profession operating in the 21st century. The planned events pose a unique and diverse set of challenges for stakeholders tasked with maintaining the health and safety of participants, transportation system safety and mobility (Latoski *et al.*, 2003, p. 2).

It should be stated that we cannot begin to look at the future without looking at the most studied event research fields and subjects today. Until three decades ago, there were no academic journals devoted to events or event management (Backman, 2018, pp. 169–170). Backman stated in his study in 2018 that one of the most prominent issues for event management in the future will be safety and security in all kinds of events. In this context, one of the key issues today is how can it evolve to help provide solutions to these concerns without making the event experience less enjoyable? It stands in front of us as just one of the questions that event organizations/managers need to answer. The main issues that are prominent and current in event studies and practice are in the role of events in ensuring the competitiveness and resiliency of a destination. In this context, security, travel restrictions/barriers and technology are very popular topics.

Safety and security concerns

Events are activities where people come together for fun, having an enjoyable time and memorable experience. People participate in events for many reasons, such as an opportunity to experience entertainment, education, culture or enrichment, or to fulfil social, commercial or political obligations.

There are a wide range of security issues, from the safety of food served at events to pandemics and terrorist attacks. It has been stated in previous studies that pandemics pose a great threat to both public health and tourism (Page and Yeoman, 2007; Scott and Gössling, 2015). One of the important issues in the successful running of an event should take into account all aspects of management, including the legal duties of the organizer. In general, these aspects include risk management, official licences and permits, general and private health and safety issues, and insurance issues (Shone and Parry, 2010, p. 171).

Any aspect of an event has an effect on the entire event. Every decision made about an event can affect how resources are allocated. Each event has several positive characteristics to exploit and obstacles to overcome. Issues such as the safety and security of the event can include uncertainty, an unknown outcome. Such ambiguous situations can include positive and negative consequences. The most effective way to deal with this uncertainty is through risk management practices. In this context, risk management is robust in ensuring that these uncertainties are properly identified, reasonably addressed, and addressed in a way that has the best chance of a beneficial outcome and the least harmful outcome. Risk management is increasingly defined as a priority in today's world so that events can be managed smoothly. Some events may be based on potentially dangerous or lucrative outcomes, and event organizers may proceed without appreciating or even understanding the obligations associated with their efforts (Silvers, 2009, p. 4). There are incidents that show that event organizers face serious problems when the safety of events is not taken seriously. For example, one of the biggest lessons learnt from the Atlanta Olympic Games is that not only is the event marred when the risk is not taken seriously enough, but there is collateral damage to other institutions and the host location (Tarlow, 2002).

Effective risk management in events involves a comprehensive and in-depth process that examines and analyses every possibility, then takes the necessary steps and allocates appropriate resources to control risks (Silvers, 2009, p. 24). All hazards and safety issues in an event are not equal and these are not easily visible. The range of possible outcomes and the effects and factors that contribute to these possible outcomes should be taken into account so that plans can be made to ensure readiness to address the consequences should the situation or outcome arise.

Risk management for events is a defined and planned process; to be effective, it must be planned and implemented consistently. It is not a one-time activity that can be completed and then finished. This process is sequential by nature, but iterative in practice (Silvers, 2009, p. 25).

Travel congestion and restrictions

A planned event is a public activity with a planned time and place that affects normal travel system operations as a result of increased visitor demand and/or reduced capacity attributed to event preparation. Following planned events held in the USA, local people stated that the events caused traffic congestion, transportation concerns and delays, and they became more sensitive to the impact of this travel congestion on their quality of life (Latoski *et al.*, 2003, p. 27).

The relationship between a pandemic and travel is central to understanding health security and global change (Burkle, 2006). One of the most important facts of pandemic processes is that travel is certainly central to epidemiology and the rate at which the virus spreads (Hon, 2013). This also means recognizing that both travel and tourism contribute to, and are significantly affected by, the spread of the disease and its economic consequences (Nicolaides *et al.*, 2019). The COVID-19 pandemic, which spread to the world at the beginning of 2020, accelerated over time and the number of people infected increased at that rate.

International, regional and local travel restrictions, introduced because of the COVID-19 pandemic, immediately affected national economies such as in tourism systems, international travel, domestic tourism, excursions and air transport, cruises, public transport, accommodation, cafés and restaurants, congresses, festivals and various events such as meetings and all sporting events. International and domestic tourism was in rapid decline for months, as international air travel slowed rapidly due to the crisis and many countries imposed travel bans, closed borders or imposed quarantine periods (Gössling *et al.*, 2020). Further, the practice of quarantine on international passengers clearly has implications for the countries' relative attractiveness to tourists once the quarantines have been eased (Jones and Burgen, 2020).

Using country population data, it was estimated that more than 90% of the world's population was in countries with some degree of international travel restrictions, and many of these countries also had some degree of restrictions on limited air travel and movement in their own countries (Gössling *et al.*, 2020). To contain the progression of the COVID-19 pandemic, many states implemented unprecedented measures, from border closures and travel bans to the suspension of visa waivers to internal mobility restrictions, including quarantines for inbound travellers, as well as a full quarantine and curfew (Seyfi *et al.*, 2020).

Quarantine processes have caused the suspension of many activities, such as the ban on people coming to visit, and social distance between people and closure of schools/universities and non-essential workplaces/workplaces, and cancellation or postponement of events (e.g. major conferences and trade shows, concerts and festivals, political debates and elections, sports seasons and the summer Olympics) (Gössling *et al.*, 2020).

Technology and social media

The year 2020 will be remembered as a year that has been redefined forever in terms of the event sector, as it is felt in all other sectors. New security concerns and the rise of virtual events seem to change the concepts of 2021 and the following years. Backman (2018, p. 170) stated that technology and social media will be at the top of the list of event management research topics.

Current technological products and developments affect today's events in many ways, for visitors: (i) to be aware of the event; (ii) to learn about what is offered by the event; and (iii) to evaluate and suggest the value of the event as a personal experience (Robertson *et al.*, 2015). Event technology is any digital tool that helps organizers plan and produce an event. Some examples are event management software, participation tools, virtual platforms, streaming services and mobile event application. As a solution to COVID-19, activity technology is evolving faster than ever. This rapid development has proved largely beneficial to the events sector; 90% of event professionals believe that technology can have a huge positive impact on the success of their event (Bizzabo, 2021). From a facial recognition app for easy registration to building an event app or using virtual reality for realistic demonstrations, digital technology can help deliver a superior experience to participants (Regalix Research, 2019).

Questions such as what event management approach should an event adopt in order to make the best use of various types of technology and especially social media sites such as Instagram, Facebook, Pinterest, and all that exists now, as well as all that will exist, are also among the questions that need to be answered today (Backman, 2018, p. 170). Mobile communications and social media are seen as expansion opportunities by digital media networks. These are tools to integrate fragmented audiences while increasing promotional circulation. For example, NBC partnered with Facebook and Instagram during both the 2012 London Olympics and the 2014 Sochi Games. Univision's adoption of Twitter during the 2014 World Cup is yet another striking example. The organizer made a deal with Twitter to participate in the sponsored tweet campaign during the tournament (Compton, 2015, p. 54). Friedman (2016) explained the following issues in his report on new trends in the event sector: (i) safety and security of events; (ii) 5G technology; (iii) artificial intelligence and bots; (iv) experiential design; (v) Internet of things (IoT); (vi) virtual reality; and (vii) augmented reality. It is thought that these issues will shape and direct the events of the future.

Sustainability and Resiliency of Events

The issue of sustainability is at the forefront of international dialogues and discussions and concerns all components of the event sector (Dickson and Arcodia, 2010). Events are profoundly linked to social, cultural values and the more basic constituent of social inclusion, a sense of belonging and a sense of identity (Goodland, 2002). The movement to create a set of frameworks for sustainable event management has resulted in a professional approach in light of contemporary concerns. It also emphasized industry best practices, advice and guidance, practical solutions, and an internal understanding of the importance of events in modern society (Musgrave and Raj, 2009).

Musgrave and Raj (2009) express the view that organizing and managing a well-planned event involves many component concerns and many stakeholders. Indicators of a good economic rationale include social and cultural benefits for a destination, raising awareness and exchange of ideas about community/social issues, networking and strengthening business contacts (Garcia, 2003).

As expressed by Cavagnaro *et al.* (2012), sustainability applied in event management is the achievement of positive impacts in the context of people, the environment and profit, with a holistic contribution to meet the economic, sociocultural and environmental needs of all stakeholders involved, including the host community. The scope of any event should include groups of participants and stakeholders involved in the event, consider social inclusion as a fundamental principle for expanding participation, and encourage interest from all aspects of the surrounding society.

Mega-events involve many dimensions, including short-lived, wide media global coverage, long-term national and international significance, and economic, social, environmental and other structures. Fayos-Solá (1998) further details the components in the sustainability of events by highlighting the multiple outcomes for the host region/country, with impacts on tourism, travel, infrastructure, social, education and other sectors. It is stated that the issues related to events and its sustainability started with the Olympic Games. In retrospect, the Olympic Games have often been found to overlook environmental and sustainability issues (Meza Talavera *et al.*, 2019). In parallel, the Norwegian authorities recognized the importance of the environment and sustainable development and embraced the sustainability concept in the 1994 Winter Games (Borcher, 2011).

Currently, it is necessary to find new solutions or new ways to solve new problems that arise in the context of event tourism. Risk management practices are at the forefront of these solutions. Risk management can be considered as the mechanism by which we seek to be aware of things that can go wrong at an event or venue, and we must make plans or take steps to prevent or mitigate these risks. These risk/safety groups are generally divided into four groups in the literature. These are:

- economic risks (e.g. the lack of financial support, sponsorship losses or lack of grants);
- performance risks (e.g. due to entertainers not being booked, an activity that does not close or lack of contestants);
- psychological risks (e.g. a place or preceding similar event that had a bad reputation); and
- physical risks (e.g. health and safety issues or crime and security challenges that pose a threat to the public or participants) (Shone and Parry, 2010, p. 171).

Risk assessment is a way of trying to identify potential risks, ensuring the resilience of the event, and taking steps to reduce or mitigate them. It is also the starting point for producing emergency plans and emergency procedures (Tarlow, 2002). Preparation in planning is crucial for the successful implementation and evaluation of any management plan. The goal of strategic thinking in sustainable incident management is to ensure that any response is timely, to achieve optimum effect, and to provide an understanding of the reason for any action to all other stakeholders, resources and resilience of events.

Conclusion and Implications

Historically and for several decades, entertainment, sporting, cultural and economic interests dominated the organization of large-scale events and games. There has been a considerable increase in interest and participation in events in the early years of the 21st century and in all corners of the world. Today, despite the negative global and regional developments, how to ensure the sustainability of these events has become one of the most important issues. Based on the event, researching, designing, planning, coordinating and evaluating activities require a balance of creative and logical thinking. The important thing for successful and contemporary governance is that these thinking processes must culminate in making good decisions, including selection, accountability and implementation. Keeping abreast of current developments and better understanding enhances this evaluation process, which helps managers make better decisions and gives more insight into stakeholders' decisions (Zrnic and Susnjar, 2010).

While events offer many opportunities, they can also include adverse risks, such as disruptions in the supply chain of important services, a decline in the quality of service, and the COVID-19 pandemic experienced today. Although the world has experienced serious health, safety and economic problems (some countries still do), over time countries have found different methods to deal with these problems. Considering the international research on event tourism in general, it is stated that events held in destinations contribute significantly to the development of tourism (Backman, 2018), and therefore activities for event tourism will continue to increase.

Events may attract large numbers of visitors in confined spaces, contain tangible goods and property of significant value, and may also be held in sensitive environments. The crowd of visitors brings with it visitor safety and security risks. Event organizers are responsible for the safety and security of the event, just as they are responsible for the pleasant experiences of the participants.

As virtual events have reached a wider audience from around the world thanks to technology, it has brought new and exciting opportunities. From software such as ticketing systems and webinar tools to innovations such as artificial intelligence and virtual reality, technology is giving signs of what an event will be like in the future (Eventbrite, 2021). With digitalization, events create a digitally connected world where they provide a combination of learning, entertainment, fun and social communication while providing a better experience for people. As event management and practices increase in less developed countries, the role of events in these countries will become even more important in planning traditional tourism strategies for the content of events, safety practices and for visitors to expand their tourism offerings to extend their stay.

References

Backman, K.F. (2018) Event management research: the focus today and in the future. *Tourism Management Perspectives* 25, 169–171.

Bizzabo (2021) Event Technology: the 2021 Guide. Available at: https://blog.bizzabo.com/event-technology-guide (accessed 15 March 2021).

Borcher, M. (2011) *Sustainable Mega-Events in Developing Countries: Experiences and Insights from Host Cities in South Africa, India and Brazil*. Konrad-Adenauer-Stiftung, Johannesburg, South Africa.

Burkle, F.M.Jr. (2006) Globalization and disasters: issues of public health, state capacity and political action. *Journal of International Affairs* 59(2), 231–265.

Business Insider (2020) UK airline Flybe declares bankruptcy as coronavirus dooms the already struggling carrier. Business Insider, 5 March. Available at: https://www.businessinsider.com/uk-airline-flybe-declares-bankruptcy-flights-grounded-2020-3 (accessed 26 October 2020).

Cavagnaro, E., Postma, A. and Neese, T. (2012) Sustainability and the events industry. In: Ferdinand, N. and Kitchin, P.J. (eds) *Events Management: an International Approach*. SAGE, Thousand Oaks, California, pp. 199–210.

Compton, J. (2015) Mega-events, media, and the integrated world of global spectacle. In: Gruneau, R. and Horne, J. (eds) *Mega-events and Globalization: Capital and Spectacle in a Changing World Order*. Routledge, Abingdon, UK, pp. 48–64.

Dickson, C. and Arcodia, C. (2010) Environmentally sustainable events: a critical review of the literature. In: *Global Events Congress IV: Festivals and Events Research: State of the Art*. Leeds Metropolitan University. 14. Available at: https://www.researchgate.net/profile/Charles-Arcodia-2/publication/265824784_Environmentally_sustainable_events_a_critical_review_of_the_literature/links/54b70c6b0cf2e68eb2800a66/Environmentally-sustainable-events-a-critical-review-of-the-literature.pdf (accessed 12 August 2021).

Eventbrite (2013) New Trends Impacting Festival and Consumer Events. Available at: https://eventbrite-s3.s3.amazonaws.com/marketing/britepapers/Britepaper_6newtrends_webVersion.pdf (accessed 25 October 2020).

Eventbrite (2021) 7 Upcoming Trends in Event Technology You Need to Know About. Available at: https://www.eventbrite.co.uk/blog/10-event-technology-trends-2020-ds00/ (accessed 26 April 2021).

Fayos-Solá, E. (1998) The impact of mega events. *Annals of Tourism Research* 25(1), 241–245.

Friedman, F. (2016) Future Trends Impacting the Exhibitions and Events Industry 2016 Update. International Association of Exhibitions and Events (IAEE). Available at: http://www.iaee.com/wp-content/uploads/2016/04/2016-IAEE-Future-Trends-Impacting-the-Exhibitions-and-Events-Industry-White-Paper.pdf (accessed 9 November 2020).

Garcia, B. (2003) Securing sustainable legacies through cultural programming in sporting events. Paper presented at The Legacy of the Olympic Games: 1984–2000. In: Centre for Olympic Studies (UAB) and Olympic Museum, Lausanne, 14–16 November 2003.

Getz, D. (2008) Event tourism definition, evolution, and research. *Tourism Management* 29(3), 403–428.

Goodland, R. (2002) *Sustainability: Human, Social, Economic and Environmental Encyclopedia of Global Environmental Change*. Wiley, Hoboken, New Jersey.

Gössling, S., Scott, D. and Hall, C.M. (2020) Pandemics, tourism and global change: a rapid assessment of COVID-19. *Journal of Sustainable Tourism* 29(1), 1–20.

Hall, C.M. (2012) Sustainable mega-events: beyond the myth of balanced approaches to mega-event sustainability. *Event Management* 16(2), 119–131.

Hon, K.L. (2013) Severe respiratory syndromes: travel history matters. *Travel Medicine and Infectious Disease* 11(5), 285–287.

Jones, M.L. (2017) *Sustainable Event Management: a Practical Guide*. Routledge, Abingdon, UK.

Jones, S. and Burgen, S. (2020) Spain to reopen to overseas tourists from July. *The Guardian*, 23 May. Available at: https://www.theguardian.com/world/2020/may/23/spain-to-reopen-to-overseas-tourists-from-july (accessed 10 December 2020).

Latoski, S.P., Dunn, W.M., Wagenblast, B., Randall, J. and Walker, M.D. (2003) *Managing Travel for Planned Special Events*. No. FHWA-OP-04-010. Federal Highway Administration (FHWA), US Department of Transportation, Washington, DC.

Martensen, A., Grønholdt, L., Bendtsen, L. and Jensen, M.J. (2007) Application of a model for the effectiveness of event marketing. *Journal of Advertising Research* 47(3), 283–301.

Meza Talavera, A., Al-Ghamdi, S.G. and Koç, M. (2019) Sustainability in mega-events: beyond Qatar 2022. *Sustainability* 11(22), 6407.

Musgrave, J. and Raj, R. (2009) Introduction to a conceptual framework for sustainable events. In: Raj, R. and Musgrave, J. (eds) *Event Management and Sustainability*. CAB International, Wallingford, UK, pp. 1–12.

Nicolaides, C., Avraam, D., Cueto-Felgueroso, L., González, M.C. and Juanes, R. (2019) Hand-hygiene mitigation strategies against global disease spreading through the air transportation network. *Risk Analysis* 40(4), 723–740.

Page, S. and Yeoman, I. (2007) How VisitScotland prepared for a flu pandemic. *Journal of Business Continuity & Emergency Planning* 1(2), 167–182.

Regalix Research (2019) Event Marketing. Available at: https://www.regalix.com/wpcontent/uploads/2019/05/CMO_Guide_Event_Marketing_2019.pdf (accessed 6 November 2020).

Robertson, M., Yeoman, I., Smith, K.A. and McMahon-Beattie, U. (2015) Technology, society, and visioning the future of music festivals. *Event Management* 19(4), 567–587.

Rojek, C. (2014) Global event management: a critique. *Leisure Studies* 33(1), 32–47. DOI: 10.1080/02614367.2012.716077.

Scott, D. and Gössling, S. (2015) What could the next 40 years hold for global tourism? *Tourism Recreation Research* 40(3), 269–285.

Seyfi, S., Hall, C.M. and Shabani, B. (2020) COVID-19 and international travel restrictions: the geopolitics of health and tourism. *Tourism Geographies* 1–17. DOI: 10.1080/14616688.2020.1833972.

Shone, A. and Parry, B. (2010) Risk management and legalities. In: Shone, A. and Parry, B. (eds) *Successful Event Management*, 3rd edn. Cengage Learning Business Press, Boston, Massachusetts, pp. 171–185.

Silvers, J.R. (2009) *Risk Management for Meetings and Events*. Routledge, Abingdon, UK.

Tarlow, P.E. (2002) *Event Risk Management and Safety*. Wiley, Chichester, UK.

Van Niekerk, M. (2017) Contemporary issues in events, festivals and destination management. *International Journal of Contemporary Hospitality Management* 29(3), 842–847.

Zrnic, M. and Susnjar, V. (2010) Risks and decision making in event management. *Communications in Dependability and Quality Management* 13(2), 19–31.

2 Managing Events and Festivals Post-COVID-19: Maintaining the Status Quo or Resetting for the Sustainability Agenda

Priyakrushna Mohanty[1]*, Anukrati Sharma[2], Jeetesh Kumar[3] and Bakhodir Turaev[4]

[1]Garden City University, Bangalore, India; [2]University of Kota, Rajasthan, India; [3]Taylor's University, Subang Jaya, Selangor, Malaysia; [4]Tashkent State Transport University, Uzbekistan

Introduction

Events tourism has often been regarded as the fastest-growing sector influencing both the demand and supply of the tourism industry (Solberg and Preuss, 2007). Apart from their primary goal of gathering economic gains of the host community, events tourism also plays an important role in building a positive image of the destination (Oklobdzija, 2015). Events are also symbolic marks of the destination because of their deep cultural association. It is in this regard, Getz (2005, 2008) applauds the power of events tourism to provide employment opportunities for the local communities while facilitating the visitors with an opportunity to experience the culture of the place they are visiting. Another significant reason for the massive success of events tourism is the changing tourist preference. The contemporary tourist of the 21st century is no longer satisfied with just visiting a destination, they want to experience its culture. This has made modern travel agents and tour agents take a keen interest in the events industry (Maguire and Hanrahan, 2017). These basic yet critical roles have made events tourism gather significant momentum in the last three decades. However, all these developments have come to a standstill since the outbreak of the biggest pandemic seen in the last century (Rosenthal, 2020).

Severe acute respiratory syndrome coronavirus 2 (SARS CoV-2) or the novel coronavirus (nCoV) was first detected in mid-December 2019 and has caused coronavirus disease 2019 (COVID-19) which has become perhaps the biggest pandemic that the modern world has experienced. Despite the sizable developments in the field of medicine and healthcare, this disease has wreaked havoc with negative implications spread to almost every country in the world (Oldekop *et al.*, 2020). With restricted travel enforced in various countries to stop the spread of the virus, economic activities have taken a major hit because of this disease. Against this backdrop, if there is an industry that has taken the biggest blow from this disease, it is the travel and tourism industry as this depends on people's mobility. As the world entered 2021, international tourist arrivals (ITAs) had reduced

*Corresponding author: pkmohanty90@gmail.com

DOI: 10.1079/9781789248685.0002

by 86% in comparison to January 2020, risking the jobs of about 100 million people employed in the tourism industry (UNWTO, 2021). Many travel-related companies were unable to cope with the financial pressure of maintaining the operational costs and went bankrupt. Subsequently, events tourism, which is an integral part of the tourism industry, also faced similar consequences. Apart from the mobilities aspect, events also involve the gathering of large numbers of people (mass gatherings) and so may increase the spread of infection. Therefore, the events tourism industry was one of the first sectors to take the hit. With strict measures such as social distancing and lockdowns that were introduced to reduce the spread of the virus, the majority of planned events were either postponed or cancelled following the COVID-19 outbreak (Mohanty *et al.*, 2020a).

Amid all these undesired developments, vaccination has come as a major relief as countries and industries are making a strong recovery (Kirksey *et al.*, 2021). Further, even though new variants of the disease are emerging over time (see Mahase, 2021) and vaccination availability and hesitancy across the world are still posing a major challenge (Chu and Liu, 2021), the economic costs of not restarting the industries are too big of a burden to carry. Therefore, as Pchejetski *et al.* (2020) note 'COVID-19 is here to stay' and humans have to learn to live with it. Against this backdrop, the tourism industry (including events), like any other industry, is gearing up for a strong recovery with various guidelines issued from the World Tourism Organization (UNWTO) and other national and multinational agencies (UNWTO, 2020).

With these positive inputs, the rebound of the events industry is very much expected to happen. But the question in this regard remains – what form should this recovery take? There have been rising voices both from the academic side and the practitioner's side to build resilience and adapt the values of sustainability to the tourism and events industry (Brouder, 2020). At the same time, the events industry has been known to avoid questions relating to sustainability in the past (Mohanty *et al.*, 2020b). These opposing realities have given rise to an interesting debate which will be discussed in this chapter. While adopting a descriptive research framework, this work provides insights into the various views associated with the future of the events tourism industry. The existing literature does not provide a detailed analysis of this diabolical situation which this chapter has tried to address. This chapter primarily takes inspiration from the qualitative data existing in the secondary literature and authors' empirical experiences.

Strength of the Events Industry

One of the fundamental building blocks of the aspect of sustainable events tourism recovery is the existing strengths of the events tourism industry. While developing a recovery plan these strengths must be reinforced and negotiated. One of the primary and perennial strengths of the events industry is its ability to make a sizable economic contribution to the local community. In particular, mega-events have been applauded for their ability to contribute greatly to the local community income, tourist expenditure, growing employment, and government revenue (Fourie and Santana-Gallego, 2011; Lee *et al.*, 2013, 2017). Magno and Dossena (2020) further stress the fact that apart from generating economic and promotional benefits for the host community, events can partially generate these benefits for the neighbouring region as well.

According to Getz and Page (2016), events are critical to a tourism system because they act as both push factors (for tourists at the originating region) and pull factors (as an attraction at the destination point). Apart from these highlighting strengths, events have created a special niche for themselves as a tool for generating positive destination image and awareness which in return provide positive impetus to local tourism (Lee *et al.*, 2005, 2017). Another positive impetus provided by the events industry is the aspect of cultural exchange (Raj, 2004). When tourists visit a particular event at any destination, they get first-hand experience of the culture of that place and when they return to their native place, they carry these cultural learnings that are assimilated with their native culture. Similarly, during their stay tourists teach some of their cultural values to the host community. These exchanges can often result in higher tolerance towards other cultural values.

Finally, events provide a competitive edge to the destinations by acting as a unique selling proposition (USP) of the destination (Dalonso *et al.*, 2014). Modern-day tourists are not satisfied with just visiting a destination, they demand experiences that are unique to the destination (Chin and Lo, 2017). Events are the best tools to satisfy these demands. Consequently, this can make the trip memorable, increasing the chances of repeat visits. All these strengths of the events industry make them one of the fastest-growing segments in the tourism sector. However, these benefits come at a cost which is discussed in the next section.

Event Vulnerabilities

The COVID-19 pandemic has been mentioned as one of the watershed moments for the whole world (Higgins-Desbiolles, 2020) with negative impacts of the disease spread to almost every economic activity (Fernandes, 2020; McKibbin and Fernando, 2020). Consequently, the global events industry, which is among one of the biggest drivers of the economy, was one of the first and worst-hit industries by the pandemic (Ahmed and Memish, 2020; Ebrahim and Memish, 2020; McCloskey *et al.*, 2020). This outcome was inevitable given the innate nature of global events that makes them much more vulnerable to external distortions such as the pandemic. Some aspects of the innate nature of events are discussed below.

Events are large systems of interconnected operations which are complex and hard to fix (Jakulin, 2020). This essentially means that because of the massive scale of the operations involved in the conduct of events, they become rigid systems that are hard to cancel or even postpone. Any small internal or external disturbance can cause a domino effect resulting in the possible collapse of the whole system. When it comes to practical implications, cancelling or postponing an event involves sending communications to all attendees and making refunds to the attendees who may not be attending the show. This can come as a huge financial burden to the event organizers and promoters as well.

Another part of the innate nature of the events industry is their higher dependence on the mobility of the people attending the events or festivals. Events are often organized in a single venue to which people from different parts of the country or the world come. The COVID-19 pandemic situation has been believed to be aggravated the most through the mobility of people, and so travel restrictions on people were put in place to control the infection rate (Seyfi and Hall, 2020). Because of this, the events and travel industries were the hardest-hit industries in this pandemic. Further, events involve the gathering of the attendees in one place (Sarwar *et al.*, 2020). This is one of the reasons why events in many parts of the world were tagged as the superspreaders of the pandemic. The case of the Kumbh festival (Quadri and Padala, 2021) and the Tablighi Jamat congregation (see BBC, 2020) in India are examples of such superspreaders.

Finally, the question of the sustainability of the events tourism industry has always clouded its future. In the last three decades, the sustainability of the events industry has been a huge question mark from all the triple bottom lines (economic, sociocultural and environmental) of sustainability. On the economic front, events have been held responsible for: (i) creating a stumbling block to local vendors' economic activities (Daniels *et al.*, 2004); (ii) facilitating only short-term results (Spilling, 1996); (iii) misusing tax-payers' money (Mules and Faulkner, 1996; Hall, 2012); and (iv) creating large-scale economic leakages (Diedering and Kwiatkowski, 2015). From the social and cultural angle, events have been blamed for staged authenticity, commodification, disruption in the quality of life of the local community, sexual assault, prostitution, increasing health disorders, and inducing vandalism and crime in the residents (Mohanty *et al.*, 2020b). Similarly, events tourism has been vehemently critiqued for its negative influences on the environment in the form of higher carbon emissions, depletion of local resources, pollution and not following green practices for earning higher profits (Getz, 2009). Besides these challenges, Mohanty *et al.* (2020b) also associate infrastructural problems such as congestion, overcrowding, traffic jams and boosting of the crowding out effect to the events tourism industry.

These major setbacks associated with events and festivals make them an easy target for any disturbances. Also, when considering the case of COVID-19, the option of ensuring peoples' health is always chosen over the recreational and economic values of the events. Events and tourism are tertiary activities and so, they are often given the last priority in case of a pandemic situation. Because of this, they are the first industries to be shut down and the last ones to be resumed in situations like COVID-19. Hence, the values of sustainability must make the events more resilient. In the next sections, some of the ways to absorb sustainability measures in the events industry are discussed in detail.

Negative Impacts of COVID-19 on the Events Industry

The impact of COVID-19 on the events industry can be termed devastating. Though there is a scarcity of data on the effect of the pandemic on events at the global level, the cancellation or postponement of almost all mega-events only suggests that the estimates of loss are expected to be massive. May it be sports, business or religion, events across the world have taken a major blow because of this pandemic. At the national and regional level, many domestic agencies have released data about their respective countries. These data sets will be crucial in devising the policies for the future.

According to a study conducted by Allied Market Research, global events accounted for a net value of US$1.1 trillion in 2019 and was expected to rise above the US$1.6 trillion mark by 2029 (Allied Market Research, 2021). However, COVID-19 has startled the whole industry. By March 2020, the events industry had incurred a loss of US$1.6 billion due to the pandemic (UFI, 2020). A recently published report by the Welsh Government in the UK provides a summary of a study of 174 event-planning organizations. The report suggests that 75% of the companies had to put their employees on furlough and the estimated median loss per company amounted to between €50,001 and €100,000 (Welsh Government, 2021). Another study conducted by EEMA (the Events and Entertainment Management Association) suggests that in India alone, the events industry which accounted for 60,000 odd events a year is expected to lose revenues amounting to ₹1 trillion with 10 million jobs at risk because of COVID-19 (Inc42, 2020). Additionally, a survey conducted by EventMB (2021) on 2500 event professionals reported that almost 90% of the professionals reported having partial or full business loss due to the pandemic. All these statistics provide a bottom line for the higher vulnerability of the events industry and the importance of building resilience. But, whether the industry is ready to learn its lessons is a matter for discussion.

For example, a study conducted by LYYTI (2020) interviewed 497 event-planning organizers of Northern Europe and found that about 50% of events were cancelled and 43% of events were postponed due to the COVID-19 pandemic. However, the same study also reported that the respondents believed that the events industry will maintain its status quo in terms of size and nature after the pandemic. While being positive, this report is also worrisome as there is no mention of sustainability in the whole report. And this is not to infer that the respondents did not care for sustainability, but the ways of recovery that the industry is planning on may often ignore the aspect of sustainability and resilience building. Also, it must be kept in mind that the nature and intensity of the impact of COVID-19 on different events vary as per their nature, size and location. For example, religious events which are impossible to postpone will be affected more than business events that have the liberty to be organized on a later date. Understanding these linkages will help in drafting better policies for the future. A detailed account of events cancelled or postponed due to COVID-19 has been provided in the work of Mohanty *et al.* (2020a).

Positive Inputs for the Events Industry from COVID-19

Despite all these adversities, many scholars and practitioners have reported many stark aspects of positive trends and developments that the pandemic may give rise to. One of the first articles to highlight the positive impacts of the COVID-19 pandemic on the events industry, in general, was that of Cooper and Alderman

(2020). Citing the example of an annual basketball event called March Madness these authors note:

> The cancellation of March Madness resulted in the loss of millions of dollars in tourism revenue for local economies and deprived traveler-fans of pilgrimages to arenas, important socio-cultural gathering spaces for American basketball fans. However, it also prevented the emission of a sizable quantity of greenhouse gasses based on our carbon footprint calculated from the previous year's tournament. Ultimately, from the disruptive closing down of sport and event tourism, a post-pandemic sports tourism landscape should emerge that takes more seriously the triple-bottom line notion of balancing a reduction of greenhouse gas emissions with the creation and maintenance of resilient local economies all while both acknowledging the important role sport plays in society and keeping tourism actors healthy.
>
> (Cooper and Alderman, 2020, p. 525)

While focusing on the positive side of the pandemic, the authors also point towards the invisible yet strong lesson of absorbing the values of sustainability that event planners may take from this pandemic. As mentioned before, the criticism of sustainability is not new to the events industry and perhaps there is no better time to start these noble initiatives to think and act upon the negative economic, sociocultural, environmental and infrastructural implications of the events industry. Similarly, Seraphin (2021) stresses the propositions developed by Getz and Page (2019) as the positive outcomes of COVID-19. Seraphin (2021) implies that due to the pandemic virtual events will be promoted which will reduce both the risk and the cost of travelling to far-away places. Similarly, it is also believed that due to the pandemic event tourism professionals will become more competent and acquire the talent of crisis management while dealing with the COVID-19 pandemic (LYYTI, 2020). Apart from this, the pandemic situation will provide a chance for the whole industry to start afresh with a new outlook and stronger motivation. Furthermore, there have been major steps taken at both individual and organizational levels to enforce the hygiene aspect in the events industry to ensure the safety and security of the attendees.

Sustainable and Resilient Events: Finding Balance between Growth vs Safety

We now get back to the primary objective of understanding what form the post-COVID-19 recovery of the events industry should take. From the above discussion, it is clear that events play a critical role in generating economic and sociocultural benefits for the host community and there must be a strong recovery for this sector to make a substantial recovery of the economy of the host community. At the same time, safety and security issues must be given priority to make this recovery fruitful. While the aggressive approach may suggest the growth of the events industry should take place by any means, the conservative approach may suggest the shutting down of events to stop the spread of infection. This may put many practitioners and policy makers in a confusing position. However, the best response to this debate, in our view, will be not choosing one of the approaches but finding the right balance between these two contradictory stances. Keeping these facts in mind, a framework for a resilient and sustainable tourism recovery plan post-COVID-19 has been developed, which has been depicted in Fig. 2.1.

The framework points to the fact that even before the pandemic, the events industry had been both commended and critiqued for its strengths and vulnerabilities, respectively. A good recovery plan will always reinforce the positive strengths and minimize the vulnerabilities to make the system sustainable. Subsequently, there have been many negative impacts and few positive inputs to the events industry because of the COVID-19 pandemic. Policy makers must appreciate these positive inputs and try to reduce and offset the negative impacts. On the whole, this will make the system even stronger than the pre-COVID event systems if this framework is effectively implemented. It must be noted that this framework is just a brief overview of the recovery plan and the contents of each element may vary from one place to another. Therefore, any modification with proper justification intended for refinement of this framework is always welcome.

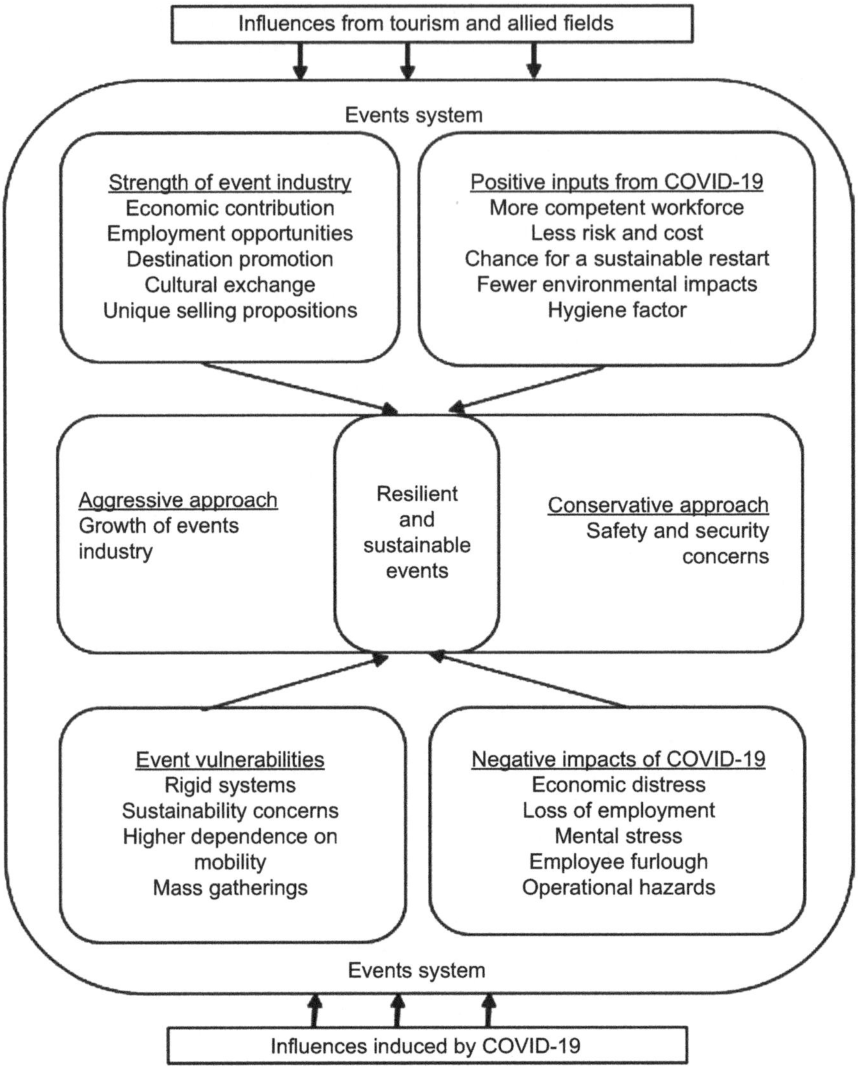

Fig. 2.1. Framework for a resilient and sustainable tourism recovery plan post-COVID-19.

Conclusion

There is absolutely no doubt that COVID-19 has had devastating impacts on the events and tourism industry. The crippled economic condition of many countries and their event industries is a devastating sight. However, with swift vaccination drives and able leadership of the national and international agencies, events tourism is prepared to launch its recovery plan. Also, as the famous quote attributed to the 19th-century German philosopher Friedrich Nietzsche states 'What doesn't kill you, makes you stronger', the events industry can gain many lessons from this pandemic which will make it stronger. The structure of the recovery plan should be such that it reinforces the classic benefits of the events industry while acknowledging the lessons learnt from this pandemic.

At the same time, efforts must be taken to minimize the vulnerabilities of the event industry while taking measures to offset the negative impacts of the pandemic effectively. A framework for such a recovery plan has been drafted, discussed and put forward in this chapter for further refinement by other researchers.

References

Ahmed, Q.A. and Memish, Z.A. (2020) The cancellation of mass gatherings (MGs)? Decision making in the time of COVID-19. *Travel Medicine and Infectious Disease* 34, 101631.

Allied Market Research (2021) Events Industry by Type. Available at: https://www.alliedmarketresearch.com/events-industry-market (accessed 13 March 2021).

BBC (2020) Coronavirus: Search for hundreds of people after Delhi prayer meeting. *BBC News*, 31 March. Available at: https://www.bbc.com/news/world-asia-india-52104753 (accessed 31 March 2021).

Brouder, P. (2020) Reset redux: possible evolutionary pathways towards the transformation of tourism in a COVID-19 world. *Tourism Geographies* 22(3), 484–490.

Chin, C.-H. and Lo, M.-C. (2017) Tourist's perceptions on man-made elements, natural elements and community support on rural tourism destination competitiveness. *International Journal of Leisure and Tourism Marketing* 5(3), 227.

Chu, H. and Liu, S. (2021) Light at the end of the tunnel: influence of vaccine availability and vaccination intention on people's consideration of the COVID-19 vaccine. *Social Science & Medicine* 286, 114315.

Cooper, J.A. and Alderman, D.H. (2020) Cancelling March Madness exposes opportunities for a more sustainable sports tourism economy. *Tourism Geographies* 22(3), 525–535.

Dalonso, Y.S., Lourenço, J.M., Remoaldo, P.C. and Netto, A.P. (2014) Tourism experience, events and public policies. *Annals of Tourism Research* 46(2), 181–184. DOI: 10.1016/j.annals.2014.03.003.

Daniels, M.J., Backman, K.F. and Backman, S.J. (2004) Supplementing event economic impact results with perspectives from host community business and opinion leaders. *Event Management* 8(3), 117–125.

Diedering, M. and Kwiatkowski, G. (2015) Economic impact of events and festivals on host regions – methods in practice & potential sources of bias. *Polish Journal of Sport and Tourism* 22(4), 247–252.

Ebrahim, S.H. and Memish, Z.A. (2020) COVID-19 – the role of mass gatherings. *Travel Medicine and Infectious Disease* 34, 191617. DOI: 10.1016/j.tmaid.2020.101617.

EventMB (2021) Your guide to successful virtual events in 2020. Available at: https://www.eventmanagerblog.com/pivot-to-virtual (accessed 14 April 2021).

Fernandes, N. (2020) Economic effects of coronavirus outbreak (COVID-19) on the world economy. *IESE Business School Working Paper* No WP-1240-E.

Fourie, J. and Santana-Gallego, M. (2011) The impact of mega-sport events on tourist arrivals. *Tourism Management* 32(6), 1364–1370.

Getz, D. (2005) *Event Management and Event Tourism*. Cognizant Communication Corporation, Elmsford, New York.

Getz, D. (2008) Event tourism: definition, evolution, and research. *Tourism Management* 29(3), 403–428. DOI: 10.1016/j.tourman.2007.07.017.

Getz, D. (2009) Policy for sustainable and responsible festivals and events: institutionalization of a new paradigm. *Journal of Policy Research in Tourism, Leisure and Events* 1(1), 61–78.

Getz, D. and Page, S.J. (2016) Progress and prospects for event tourism research. *Tourism Management* 52, 593–631.

Getz, D. and Page, S.J. (2019) *Event Studies: Theory, Research and Policy for Planned Events*. Routledge, Abingdon, UK.

Hall, C.M. (2012) Sustainable mega-events: beyond the myth of balanced approaches to mega-event sustainability. *Event Management* 16(2), 119–131.

Higgins-Desbiolles, F. (2020) Socialising tourism for social and ecological justice after COVID-19. *Tourism Geographies* 22(3), 610–623.

Inc42 (2020) How Events Industry in India is Tackling Impact of Covid-19? *Inc42*, 10 May. Available at: https://inc42.com/resources/the-impact-of-covid-19-on-the-events-industry/ (accessed 14 April 2021).

Jakulin, T.J. (2020) Systems approach to cultural tourism and events. *Academica Turistica – Tourism and Innovation Journal* 12(2), 185–191.

Kirksey, L., Sorour, A.A. and Modlin, C. (2021) COVID-19 vaccination: potential challenges and reforms. *Journal of the National Medical Association* 113(4), 368–370.

Lee, C.-K., Lee, Y.-K. and Lee, B. (2005) Korea's destination image formed by the 2002 World Cup. *Annals of Tourism Research* 32(4), 839–858.

Lee, C.-K., Kang, S.K. and Lee, Y.-K. (2013) Segmentation of mega event motivation: the case of Expo 2010 Shanghai China. *Asia Pacific Journal of Tourism Research* 18(6), 637–660.

Lee, C.-K., Mjelde, J.W. and Kwon, Y.J. (2017) Estimating the economic impact of a mega-event on host and neighbouring regions. *Leisure Studies* 36(1), 138–152.

LYYTI (2020) The State of Events – How will COVID-19 change the event industry. Report. Available at: https://mb.cision.com/Main/19963/3238703/1336234.pdf (accessed 16 April 2021).

Magno, F. and Dossena, G. (2020) Pride of being part of a host community? medium-term effects of mega-events on citizen quality of life: the case of the world expo 2015 in milan. *Journal of Destination Marketing & Management* 15, 100410. DOI: 10.1016/j.jdmm.2020.100410.

Maguire, K. and Hanrahan, J. (2017) Assessing the economic impact of event management in Ireland: a local authority planning perspective. *Event Management* 21(3), 333–346. DOI: 10.3727/152599517X14942648527545.

Mahase, E. (2021) Covid-19: what new variants are emerging and how are they being investigated? *BMJ* 372, 158. DOI: 10.1136/bmj.n158.

McCloskey, B., Zumla, A., Ippolito, G., Blumberg, L., Arbon, P. *et al.* (2020) Mass gathering events and reducing further global spread of COVID-19: a political and public health dilemma. *The Lancet* 395(10230), 1096–1099. DOI: 10.1016/S0140-6736(20)30681-4.

McKibbin, W.J. and Fernando, R. (2020) The global macroeconomic impacts of COVID-19: seven scenarios. *CAMA Working Paper* No.19/2020.

Mohanty, P., Dhoundiyal, H. and Choudhury, R. (2020a) Events tourism in the eye of the COVID-19 storm: impacts and implications. In: Arora, S. and Sharma, A. (eds) *Event Tourism in Asian Countries: Challenges and Prospects*. Apple Academic Press, Burlington, Ontario, Canada, pp. 97–114.

Mohanty, P., Singhania, O. and Hasana, U. (2020b) Mega-events tourism and sustainability: a critique. In: Arora, S. and Sharma, A. (eds) *Event Tourism in Asian Countries: Challenges and Prospects*. Apple Academic Press, Burlington, Ontario, Canada, pp. 219–234.

Mules, T. and Faulkner, B. (1996) An economic perspective on special events. *Tourism Economics* 2(2), 107–117. DOI: 10.1177/135481669600200201.

Oklobdzija, S. (2015) The role of events in tourism development. *Bizinfo Blace* 6(2), 83–97. DOI: 10.5937/BIZINFO1502083O.

Oldekop, J.A., Horner, R., Hulme, D., Adhikari, R., Agarwal, B. *et al.* (2020) COVID-19 and the case for global development. *World Development* 134, 105044. DOI: 10.1016/j.worlddev.2020.105044.

Pchejetski, D., Atkins, J. and Lall, R. (2020) COVID-19 is here to stay, what can we do? *British Journal of General Practice*. 18 July. Available at: https://bjgp.org/content/covid-19-here-stay-what-can-we-do (accessed 18 April 2021).

Quadri, S.A. and Padala, P.R. (2021) An aspect of kumbh mela massive gathering and COVID-19. *Current Tropical Medicine Reports* 29, 1–6. DOI: 10.1007/s40475-021-00238-1.

Raj, R. (2004) The impact of cultural festivals on tourism. *Tourism Today* 4(1), 66–77.

Rosenthal, M. (2020) Fauci: COVID-19 Worst Pandemic in 100 Years. *Infectious Disease Special Edition (IDSE)* 21 October. Available at: https://www.idse.net/Covid-19/Article/10-20/Fauci--COVID-19-Worst-Pandemic-in-100-Years/60937 (accessed 18 April 2021).

Sarwar, F., Panatik, S.A. and Jameel, H.T. (2020) Understanding public's adoption of preventive behavior during COVID-19 pandemic using health belief model: role of appraisals and psychological capital. *Advance Social Sciences & Humanities* preprint posted on 17 July.

Seraphin, H. (2021) COVID-19: an opportunity to review existing grounded theories in event studies. *Journal of Convention & Event Tourism* 22(1), 3–35. DOI: 10.1080/15470148.2020.1776657.

Seyfi, S. and Hall, C.M. (2020) Sanctions and tourism: conceptualisation and implications for destination marketing and management. *Journal of Destination Marketing & Management* 15, 100381.

Solberg, H.A. and Preuss, H. (2007) Major sport events and long-term tourism impacts. *Journal of Sport Management* 21(2), 213–234. DOI: 10.1123/jsm.21.2.213.

Spilling, O.R. (1996) Mega event as strategy for regional development: the case of the 1994 Lillehammer Winter Olympics. *Entrepreneurship & Regional Development* 8(4), 321–344.

UFI (2020) Media Release: UFI releases first global assessment of the economic impact that COVID-19 is having on trade shows and exhibition. UFI – The Global Association of the Exhibition Industry, 10 March. Available at: https://www.ufi.org/wp-content/uploads/2020/03/UFI_media_release_economic_impact_of_covid19_100320.pdf (accessed 16 April 2021).

UNWTO (World Tourism Organization) (2020) UNWTO Launches Global Guidelines to Restart Tourism. UNWTO, 28 May. Available at: https://www.unwto.org/news/unwto-launches-global-guidelines-to-restart-tourism (accessed 3 June 2021).

UNWTO (World Tourism Organization) (2021) International Tourism and COVID-19. Available at: https://www.unwto.org/international-tourism-and-covid-19 (accessed 3 June 2021).

Welsh Government (2021) Events industry and supply chain business COVID-19 impact survey. Available at: https://gov.wales/events-industry-and-supply-chain-business-covid-19-impact-survey-12-november-8-december-2020-html (accessed 3 June 2021).

3 Parang Music as an Attraction for Rural Development: an Example from the Village of Lopinot, Trinidad

Johnny Coomansingh*
Independent Researcher/Cultural Geographer, Newark, Delaware, USA

Introduction

To give more clarity and salience to this chapter, a brief history of Trinidad as it relates to parang (parranda) will be entertained. In the annals of history, it is well known that Christopher Columbus rediscovered the island of Trinidad on 31 July 1498 (Watson, 1979). Although there is a bit of uncertainty about what the island was called back then, whether it were the Amerindian names, *Kairi*, *Caeri* or *Iere* (Laurence, 1967), Columbus on his arrival christened the island *La Isla de la Trinidad*, which meant *The Island of the Trinity*. This is now shortened to Trinidad. The mere mention of the word *Trinity* conveyed the idea that the Roman Catholic religion was immediately implanted on the island. This fact about the name of Trinidad is important because it confirms that the Roman Catholic influence began at the island's rediscovery. Subsequent to the arrival of the Spanish, who Knight (1990) described as restless and bellicose, the hapless indigenous population resident on the island was decimated to a mere handful, falling 'victims to European brought diseases among which was venereal disease ...' (Watson, 1979, p. 23).

The offspring of a few, particularly Caribs, have survived to this day, who inadvertently found themselves involved with Spanish parang music. According to Ingram (2008, p. 65), 'The myth of the total decimation of Amerindians in Trinidad has been countered by the organization and revival of a group called the Santa Rosa Carib Community (SRCC) located in the central town of Arima.' In the annual festival of the SRCC, staged jointly with the Santa Rosa Catholic Church of Arima, parang music with origins in Venezuela became a highly noticeable part of the festivities (Ingram, 2008).

Parang is essentially a Spanish Roman Catholic musical artform. Without question, Trinidad became a Spanish colony that never saw a moment of rest. The British, Dutch and French conducted raids from its inception throughout the 18th century. In essence, Trinidad's development was orchestrated by influences from at least four major European countries. Nonetheless, the Spanish held on to Trinidad until the British overthrew them in the year 1797. Spanish colonization concomitant with the enslavement of the Amerindian population saw the cultivation of sugarcane, tobacco and animal husbandry on a small scale. During this era, Trinidad was basically a jumping off point to the South American mainland. The South

*nanaba7@gmail.com

DOI: 10.1079/9781789248685.0003

American mainland proved to possess deposits of precious metals and gems. Actually, from the mid-1700s, Trinidad was under the economic control of the French, the French having been ceded lands by the Spanish (Williams, 1964; Hill, 1972).

After the French, the impact of the British taking control from the Spanish was shocking. 'The British seized Trinidad in 1797 ...' (Watson, 1979, p. 90) then bullied their way in the colony and foisted Anglicanism on the predominantly Roman Catholic population. This was a clash of religions on the landscape; Trinidad was now a contested religious landscape. In essence, Mitchell (2000, p. 109) provided an explanation for such a landscape by stating: 'Landscape, and landscape representations, are thus incorporations of power ... They are made to actively *represent* who has power, certainly, but also to *reinforce* that power by creating a constant and unrelenting symbol of it.' Trinidad eventually became a crown colony and went full scale into the production of sugar (Watson, 1979). Nevertheless, the cocoa and coffee industries in Trinidad survived under French control and absorbed labour from the Spanish *peons* from Venezuela, thus transferring their Spanish culture to Trinidad, the parang artform being more prominent. Therefore, the parranda as a Roman Catholic institution was an indirect means to cement Roman Catholicism as the dominant religion in the colony through song, music and gastronomic preferences. Parranda as both a physical and a religious force of Roman Catholicism helped in no small way to retain Roman Catholicism as the dominant religion in Trinidad to this day. The research aims to answer the question: Why was the village of Lopinot the site chosen for *La Fiesta de Lopinot?*

Study Area

The area chosen for study is the village of Lopinot, 'the most elevated village in Trinidad' (NALIS, 2020a). Lopinot is basically a miniature *altiplano*. About 9 km (5.59 miles) from the Eastern Main Road, north of Arouca, the closest town to the village, Lopinot is nestled in a lush valley situated at the foothills of the Northern Range. With a population of about 3000 people, Lopinot exhibits a complex mixture of descendants from Spanish (*Cocoa Panyols*), French, Portuguese, Chinese, West African, East Indian and Amerindian (Carib), most of whom follow the Roman Catholic way of life. In terms of language, the villagers speak the English language (formal and informal), a bit of Spanish and Patois (broken French). In fact, the *paranderos* sing their songs primarily in the Spanish language. Despite the plural nature (racial mixture) of the residents, everyone is directly or indirectly involved in parang in the village of Lopinot.

The village with its agrarian lifestyle was labelled Lopinot due to the conquest of the valley by Charles Joseph Comte de Lopinot, a powerful planter and former Knight of the Royal and Military Order of St Louis. Count Lopinot tried his hand at the cultivation of sugarcane but soon realized that the soils were not suitable for sugarcane production. With the help of 100 enslaved Africans, he converted dense tropical forests to cocoa plantations.

He named his newly acquired estate of 478 acres *La Reconnaissance*. Today the bustling little village, complete with shops, restaurants, guest houses, parlours, rum shops (pubs/drinking places), schools, churches and a community centre, provides spaces for villagers to express themselves. These locations afford opportunities for people to meet, greet, plan, play and discuss village matters (Homer, 2013). It is important to note also that the Lopinot River flows adjacent to the Lopinot Complex, a place where many people from all over Trinidad come to relax and enjoy themselves in the crystal-clear water. According to Homer:

> The spectacularly remote, cute village of Lopinot is home to year-round parang music ... the weekend is the best time to take your family to Lopinot village for clean air and a relaxing river *lime* (lime—hangout or shoot the breeze) under the grand old trees.
>
> (Homer, 2013)

The grand old trees are Samaan trees (*Samanea saman*) or rain trees that complement the landscape of what has been named the Lopinot Complex.

Among the persons of note who once existed in the village are Knolly Gomez, Sotero Gomez, Domingo Gomez, Pedro Segundo Dolabaille (Papa Goon) and Leonard Burnley, who became known for their cultural exploits especially in the arena of parang music (Baboolal, 2008). Still living today is Martin Gomez, a parandero and Lopinot historian. It is said that Lopinot has its fair share of superstition. The idea that the Lopinot estate house is haunted by a ghost has gone on record by ghost hunters. Note that 'in 2011, Ghost Hunters International and the SyFy Channel featured the site on its show for paranormal activity around the legend of the Count' (Chaitram, 2021). It is believed that the Count tortured, ill-treated and murdered some of his slaves, thus the legend lives that he is yet to find a resting place. A cashew tree growing near to the cocoa drying house is believed to be 'The Hanging Tree' where enslaved Africans were hung when the Count considered them to be useless (Chaitram, 2021).

Quite noticeable is an icon of superstition that reminds us of the idea that the *evil* eye is alive and well in Lopinot because of the presence of upside-down bluebottles fixed to bamboo poles. These poles stand erect in the vegetable gardens and surroundings of the villagers. The bluebottles are said to ward off the evil eye or *maljo* (from French mal yeux). Many people entertain the notion that the bluebottles will also keep away thieves. There is the belief among some people that if they steal the crops and eat them, they will suffer with 'swell belly' (Homer, 2013). Lopinot also boasts of the lure of an exciting karst landscape in the explorable caves that exist in the area. For all wants and purposes, Lopinot is a wonderful tourist paradise, a little agricultural village wrapped up in the traditions of the past, holding together a mixture of peoples who love life and the sweet melodies of parang music.

What is Parang?

The term **parang** is derived from the Spanish word *parranda*, which means a spree or a fete. Initially it meant a group of four or more men who went to give a *parranda* at an event – a christening or a birthday celebration. The group sang to the accompaniment of musical instruments. However, in Trinidad, parang came to mean the songs that were sung especially during the Christmas season. What was brought from Venezuela to Trinidad was *parranda navideña*, which means *Christmas parang* (Coomansingh, 2019). Taylor and Thomas (1977) pointed out that parang is essentially a part of Trinidad's heritage. The word *parang* is neither Spanish nor English but represents the idiom for *parran*, which is a shortened form of *parranda*. Parranda in the truest sense involves a group of people, sometimes four or five and maybe more, going from house to house playing musical instruments and singing, carousing and feting. Of course, eating and drinking are integral to such partying.

There are two theories about the origins of Trinidad parang. The first is that the custom was brought to the island by Capuchin missionaries during the Spanish occupation of Trinidad from 1498 to 1797. The parang musical artform continued to flourish even after the British took over the island. The British could not stop the constant interaction between the people of Trinidad and those of Venezuela (the Spanish mainland). The second theory suggests that the custom was brought over from Venezuela in the 19th century by the *Cocoa Panyols* who came from Oriente, east Venezuela, to work on the cocoa plantations in Trinidad. Whatever its origins, the *infectious* parang came as part of the *cultural baggage* (Zelinsky, 1992) of the Spanish. The artform is now an integral part of the cultural landscape of Trinidad and Tobago. Incorporating aspects of indigenous and South American cultures, parang has been called a fusion of 'the deep spiritual aspirations of the Spanish people and the unfettered joyfulness of the Amerindian and African cultures' (Coomansingh, 2019, p. 10).

From what I have witnessed, sometimes many paranderos go paranging with just a cuatro, a pair of maracas, a tock-tock and sometimes a box bass. This type of social paranging occurs in several recognized places in Trinidad. Some of the more important towns and villages include: Arima, Caura, Lopinot, Mausica, Palo Seco, Rio Claro, Saint Ann's, Saint Joseph, San Raphael, Santa Cruz, Sangre Grande, Siparia and Tamana. The official parang season begins on 25 September and continues to 6 January.

Various parang groups participate in competitions organized by the National Parang Association of Trinidad and Tobago (NPATT) during this period. Nonetheless, there are paranderos who parang way past the closing date. Some years ago, in a little chat with a taxi driver whom I came to know only as 'Mr Vialva', I discovered that he was indeed an obsessive listener of parang music. Plying his trade in the town of Sangre Grande, he plays parang music all day, every day, in his taxi. In Trinidad, it is well known that parang culminates on 6 January 'The Day of the Kings or *Dia de los Reyes* (known elsewhere as Epiphany)/The Feast of Kings' otherwise known as *Lewah* (Les Rois).

Parang Instruments

There are a variety of instruments used to accompany singers in parang music. However, the cuatro, tock-tock, guitar maracas (chac-chac) and the box bass seem to be the instruments most commonly used today. Because the heritage of Trinidad is so varied, many parang bands may use different instruments including the steelpan instrument, which was invented in Trinidad (Coomansingh, 2005, 2018). In this regard, European, African and South Asian influences come to the fore. Even though parang in Trinidad is deeply rooted in Spanish culture, some individuals have used the *dholak*, a South Asian drum, in a parang session in the village of Tamana where the residents are a mixture of mainly Spanish, African and South Asian influences. According to the Trinidad and Tobago National Library and Information System Authority (NALIS) website (2020b) 'it is now difficult to identify one group of instruments as the core of the parang band'. However, most bands would have some combination of the following instruments: cuatro, mandolin, tiple, violin, bandol (bandola), bandolin, guitar, box bass, maracas ('chac-chac'), wood block (claves or 'tock-tock'). Contemporary bands have included other instruments such as flute, scratcher (güiro), electric bass, Latin percussion and steelpan instruments.

Let's Go Parang

Knowing the songs, instruments and musical rhythms involved in parang music is not that difficult. Nevertheless, without some of the instruments and lyrical content, people make up all kinds of tunes, sometimes combining *calypso* and *chutney* melodies. In such a fete, participants will use anything that comes to hand to make a 'noise', for example a bottle and spoon. Taylor and Thomas explain:

> The local revelry that is parang is the visiting of merrymakers to the homes of family, friends or patrons to sing songs in *Spanish* to the accompaniment of certain musical instruments; usually, the guitar, cuatro, maracas (chac-chacs), mandolin, bandolin, violin bandola and sometimes the cello.
>
> (Taylor and Thomas, 1977, p. 15)

Eating, drinking and dancing are also included in the festivity.

In support of Taylor and Thomas (1977), Moodie-Kublalsingh (1994) adds that essentially, Trinidad *parang* is the same Spanish *parranda* used primarily in *andar de parranda* – merrymaking, carousing; to go paranging with friends, which is very similar to the Venezuelan *parrandear*. Parrandear involves an atmosphere of *liming* with the movement of a group of musicians and singers going about from place to place without any time limit in mind. (*Liming* in Trinidad refers to 'shooting the breeze' or just hanging out.)

According to Festa (2013), liming is integral to the country's culture; food, drink, conversation, especially the sharing of jokes, will all combine at times to create a lime. The word came into vogue because of British sailors who hung about while their ship was being loaded with supplies, especially limes (the juice of which helped them to avoid scurvy while at sea). So the idea of paranging is almost synonymous with liming. Ingram (2002) also conveys the idea that paranging involves merrymaking, dancing, music, drinking and eating foods created by the Spanish Creole community of Trinidad. The artform could be performed anytime of the year in other festivities and celebrations, but the true signature of parang on the landscape of Trinidad is at Christmas time.

In terms of the gastronomic preferences for the parang season, there will be a lacklustre session if the regular parang 'food' is not served at the fete. Ingram (2002) related that the foods typically found at a parang event are also traditional Christmas fare to most citizens who celebrate Christmas. Such foods include: (i) *pastelles* (seasoned steamed ground chicken, beef or pork-filled cornmeal patties wrapped in banana leaves); (ii) *empanadas* (fried version of pastelle); (iii) *paime* (sweet pastelle made with corn and grated coconut); and (iv) *sorrel* (*Hibiscus sabdariffa*) a spicy cider, ginger beer, rum, *ponche de crème* (a type of alcoholic eggnog) and '*babash*' (home-brewed alcohol). Local wild meats are also prepared. Brocket deer (*Mazama americana*), agouti (*Dasyprocta leporina*), lappe (*Cuniculus paca*), tattoo or armadillo (*Dasypus Linnaeus*) and iguana (*Iguana iguana*) are some of the preferred delicacies. Servings at more intimate family gatherings may include homemade bread and ham, roast pork, pelau (savoury chicken, pigeon peas and rice cook up), stewed common fowl or stewed pork over ground provisions such as cassava, dasheen and yam. Sometimes cassava bread (an Amerindian tradition) is served (Ingram, 2002). With this array of music and food preparations in view, Lopinot seized the opportunity to stage an annual festival: *La Fiesta de Lopinot*.

Enhancing the Economy of Lopinot

Apart from the involvement with crop husbandry and the raising of a little livestock, Lopinot has looked at other means for enhancing the sustainable economic status of the community. According to the findings of the Caribbean Natural Resources Institute (CANARI) workshop held in Lopinot in October 2010, *Improving livelihoods in rural communities in Trinidad and Tobago by developing small business ideas based on the sustainable use of natural resources*, it was found that Lopinot has a wealth of natural resources to offer. Lopinot natural resources, existing uses, potential uses and water resources include:

- the watershed, river, riverside savannah – the Water and Sewerage Authority (WASA) maintains the supply of water to the community and also the river is used for riverside excursions, cookouts and bathing;
- recreation, as a tourist attraction, for ecotourism (forest trails and caves need to be maintained);
- agricultural lands and government forest reserves for:
  - production and agro-processing of food crops for subsistence and sale in the market – crops include tomato, sweet pepper, ground provisions (yam, dasheen, sweet potato, banana, plantains, etc.);
  - watermelon – whole fruit and juice;
  - maize (corn) – corn flour, boiled and roast corn for visitors;
  - plantain and other ground provisions to provide plantain chips, cassava chips, etc.;
  - mangoes – small-scale selling of mangoes, mango preserves (e.g. mango jam) to supply the grocery shop, juices, export of mangoes, cosmetics (with mangoes);
  - sour cherries – preserves and juice;
  - forests – most of the forests are in government reserves; the caves offer a great tourist attraction;
  - forest flowers – including balisier (*Heliconia bihai*)/other flowers (balisier is used as the symbol to represent the People's National Movement political party of Trinidad and Tobago), flowers used for domestic decor, exotic flower industry, healing oils such as Carapa oil (*Carapa guianensis*) and Zebapique (*Neurolaena lobata*);
- coffee and citrus – local coffee and citrus processing;
- cocoa and coffee estates (private, but of public use) provide local cocoa and coffee products and also:
  - cocoa drying house and activities related to cocoa drying, for example 'Dancing the Cocoa' (tourist attraction);
  - historical aspects/storytelling;
- birds – parrots and other species for birdwatching; and
- wildlife (e.g. agouti, lappe, manicou) – hunting for domestic use and sale.

Having a better knowledge about what the community has and what they could do with what they have and secondly, how they could use parang, the research beckons to *La Fiesta de Lopinot* and the reasons why this fiesta was inaugurated. The Lopinot Tourism Association (LTA) saw the need for enhancing the economic position of their community. Staging an annual fiesta seemed to be a great idea due to the fact that many of the residents were already involved in the making of traditional Christmas food and drinks, as mentioned above. On Saturday 13 December 2008, Martin Gomez, a parandero who is a member of the Living Legends of Lopinot, spoke with Yvonne Baboolal, a reporter from the *Trinidad & Tobago Guardian* newspaper. He said that nobody will open their door for a parang group if they come with the attitude to drink rum. The parang group must come with the 'real thing'. They must let the household know that they have come to bring greetings and the good tidings that Christ has come.

This is what the village of Lopinot aims to foster at *La Fiesta de Lopinot*: to keep the parang traditions alive. The fiesta uses the parang artform to their advantage as an attraction for citizens far and wide to come to Lopinot for the real thing. Encouraged by Roman Catholic culture, the true essence of parang is totally caught up in the celebration of the Christ child. Every traditional parang song celebrates none other than the Virgin Mary and the arrival of the Christ child. So it is more than just song, music, dance, food and feting. It goes deeper into a celebration of a religious concept that arrived on the landscape about five centuries ago.

For what it's worth, Lopinot started *La Fiesta de Lopinot* in the year 2007. And as most Trinidadians (Trinis) would say, 'rain or shine, thunder or lightning', hundreds of people, including Venezuelans and Colombians, make the 'pilgrimage' to that village to join in the festivities, bringing with them much needed cash. Under their various tents decorated with brightly coloured 'Christmas' bulbs and banners, concessionaires sometimes make up to TTD10,000 (about US$1500) on fiesta day. Ten or 12 parang bands will be in attendance keeping the day bright with music everywhere. Some bands will stop and perform on the street adjacent to the Lopinot Complex much to the joy of the people who show up to dance and have a good time. According to Donna Mora, former president of the LTA, pork is the meat that is sold the most during the fiesta. An all-time traditional favourite of the Lopinot people is *roucou pork*, which is pork coloured with colouring made from the seeds of the annatto or achiote (*Bixa orellana*) tree (Baboolal, 2008). This means that there is an opportunity for animal farmers in the village to raise many hogs (pigs) for this occasion. Apart from food and drink, many other products are also made and sold, for example cocoa and coffee products and handicraft.

Conclusion

Giving attention to some comments from Henry-Kunzel (1994), it is a well-known fact that Trinidad and Tobago did not view tourism as a serious factor in the development of the country because of the fortunes that had been made from hydrocarbon exploitation and exploration (i.e. the view was that there was no economic need for tourism). So despite the best efforts of the LTA to create opportunities for betterment in the community, there have been factors that veer toward possible failure. According to Baboolal (2008): 'A stalwart in the community of Lopinot is fearful that the culture and traditions of the area may be soon lost if government and stakeholders continue to fail to act in revitalizing its heritage and its famous Lopinot Historical Complex.' Martin Gomez, former president of the village council, spoke firmly about the dire need for government's involvement to save the Lopinot Complex. Gomez also expressed his dissatisfaction concerning the road to Lopinot. Poor maintenance of the long winding mountainous road to Lopinot can stymy the efforts of people who want to visit the village. Trees and bamboo stalks are always hanging over the roadway. Landslips and potholes are always present along the route. Gomez pointed out that the government must take a vested interest in the village of Lopinot because of its place in the history of Trinidad (Baboolal, 2008).

Gomez, who is a seasoned parang man and cuatro player, informed about the parang sessions he knew of yesteryear. He said in Baboolal (2008) that the art of paranging began long before Christmas. As noted before, NPATT declared 25 September as the date for the commencement of parang. Sometimes the paranderos begin on Christmas Eve and they go 'right down to the morning and spend Christmas Day paranging different houses' (Baboolal, 2008).

Gomez informed that he learnt to play the cuatro just like that. His father, Knolly Gomez, and uncles, Sotero and Domingo Gomez, were instrumental in his development as a parang man. He recalled that in those days Lopinot paranderos of some esteem including Pedro Segundo Dolabaille, otherwise known as Papa Goon, composed their own parang songs. Leonard Burnley, their instrument maker, eventually passed. With the passing of Burnley, the tradition of making parang instruments in Lopinot faded. Burnley's grandson took up the trade, but for whatever reasons, stopped production. Vanishing from the scene in Lopinot are parang instrument makers and composers of Spanish tunes, but that's not all.

Over a decade ago, Lopinot House remained closed to visitors. Lack of much-needed funding was the reason for the delay in repair and reconstruction works. Donna Mora, Martin Gomez's daughter, also gave her reasons why things have gone awry in Lopinot. She said: 'it was not just a lack of funding but internal conflict within the community that has also contributed to the downfall of Lopinot. It's been four years now since *La Fiesta de Lopinot* was staged' (Mora cited in Chaitram, 2021). Exacerbated by coronavirus disease 2019 (COVID-19) protocols, who knows when the fiesta will be revitalized. Mora lamented:

> We are known for our parang and this has not happened in the last four years because there isn't any camaraderie anymore. Everything has fallen apart. We have a complicated community where everybody is related in one way or the other. I think the pulling and tugging internally are the reasons for the failed heritage festival over the years.
>
> (Mora cited in Chaitram, 2021)

She called for the controlled use of and visit to the facility through appointments and the implementation of a fee structure so that funding can be generated to assist with its upkeep and other expenses (Chaitram, 2021).

Nevertheless, according to McKenzie (2021), the Lopinot Community Center situated a few metres away received a much-needed facelift through a monetary grant of TTD50,000 from the Digicel Group, an international mobile phone network and home entertainment provider. It is noteworthy to add that the LTA submitted its application to Digicel in November 2020 with 100 other applicants. The LTA and four other organizations got grants the following December.

With this money, the LTA was able to repaint the entire community centre, refurbish all the toilets, and build a new stage. As a community-based project, the LTA executed all the works and refurbishment of the centre. McKenzie (2021) added: 'The project is also expected to provide sustainable employment for young parranderos and budding chocolatiers and, as such, the group purchased parang instruments and equipment needed for chocolate making.' With this achievement, Mora's despondent feeling about 'internal conflicts within the community' were short-lived. The founder of the LTA was exuberant when she stated: 'We want the world to know about Lopinot ... The residents who live here are passionate and friendly ... We are happy to show Trinidad and Tobago that the residents of Lopinot can work together' (McKenzie, 2021).

So in view of the present state of affairs with *La Fiesta de Lopinot*, one only has to wait until the COVID-19 pandemic protocols change. Presently, all beaches, rivers, parks and recreation sites are closed. Supermarkets and pharmacies are the only businesses that are mandated to remain open. This situation has wreaked havoc on the volume of visitors coming to Lopinot. However, with the way things are going, change will not be too soon. Nevertheless, it is hoped that the Ministry of Tourism will endeavour to become totally involved in the Lopinot Complex. This comes with the attendant adequate funding to augment the plans and purposes of the village of Lopinot for a better community thus ensuring the annual staging of *La Fiesta de Lopinot*. The continuance of *La Fiesta de Lopinot* will surely enhance the perpetuation of the parang artform not only to survive, but to thrive.

References

Baboolal, Y. (2008) Paranging in Lopinot. Trinidad & Tobago Guardian, 13 December. Available at: https://www.guardian.co.tt/article-6.2.317177.f6878bf894 (accessed 28 January 2021).

Caribbean Natural Resources Institute (CANARI) (2010) *Improving livelihoods in rural communities in Trinidad and Tobago by developing small business ideas based on the sustainable use of natural resources. Lopinot, Trinidad*, 20 October. Available at: https://canari.org/wp-content/uploads/2014/09/JBFMTLopinot1.pdf (accessed 29 August 2020).

Chaitram, R. (2021) Famed Lopinot House falls to ruins, MP vows to restore landmark. *Trinidad and Tobago Newsday*, 3 January. Available at: https://newsday.co.tt/2021/01/03/famed-lopinot-house-falls-to-ruins-mp-vows-to-restore-landmark/ (accessed 31 January 2021).

Coomansingh, J. (2005) The commodification and distribution of the steelpan as a conflicted tourism resource. PhD dissertation, Kansas State University, Manhattan, Kansas.

Coomansingh, J. (2018) Authenticity of a national icon: the Trinidad steelpan as a tourism resource. In: Rickly, J.M. and Vidon, E.S. (eds) *Authenticity and Tourism: Materialities, Perceptions, Experiences*. Emerald Publishing, Bingley, UK, pp. 89–107.

Coomansingh, J. (2019) *Cocoa Woman*. Toplink Publishing, Kelly, North Carolina.

Festa, J. (2013) Carefree Caribbean: What is Liming? *Epicure & Culture: Food, Wine & Culture for the Ethical Traveler*. Available at: https://epicureandculture.com/what-is-liming/ (accessed 31 January 2021).

Henry-Kunzel, G. (1994) Tobago expects tourism to spur economy. *Hotel and Motel Management* 209(10), 21.

Hill, E. (1972) *The Trinidad Carnival: Mandate for a National Theater*. University of Texas Press, Austin, Texas.

Homer, L.B. (2013) Lopinot Village Trinidad. *The Trinidad Express*. Available at: http://www.napturalborntraveller.com/lopinot-village-trinidad/ (accessed 29 August 2020).

Ingram, A.K. (2002) *What is Parang?* Available at: http://aingram.web.wesleyan.edu/parangdescription.html (accessed 31 January 2021).

Ingram, A.K. (2008) Reading history, performing Carib: the Santa Rosa Festival and Amerindian identity in Trinidad. *Caribbean Studies* 36(2), 65–94.

Knight, F. (1990) *The Caribbean: the Genesis of a Fragmented Nationalism*. Oxford University Press, Oxford, UK.

Laurence, K.M. (1967) Notes on Iere, the Amerindian name for Trinidad. *Caribbean Quarterly* 13(3), 45–51.

McKenzie, R. (2021) Lopinot Community Centre refurbished with Digicel's help. *Trinidad and Tobago Newsday*, 17 April. Available at: https://newsday.co.tt/2021/04/17/lopinot-community-centre-refurbished-with-digicels-help/ (accessed 19 April 2021).

Mitchell, D. (2000) *Cultural Geography: a Critical Introduction*. Blackwell Publishers, Oxford, UK.

Moodie-Kublalsingh, S. (1994) *The Cocoa Panyols of Trinidad: an Oral Record*. Academic Press, London.

NALIS (Trinidad and Tobago National Library and Information System Authority) (2020a) Lopinot. Available at: www.nalis.gov.tt (accessed 29 August 2020).

NALIS (Trinidad and Tobago National Library and Information System Authority) (2020b) Parang. Available at: https://www.nalis.gov.tt/Resources/Subject-Guide/Parang (accessed 3 September 2020).

Taylor, D.P. and Thomas, G.A. (1977) *Parang of Trinidad*. National Cultural Council of Trinidad and Tobago, Port of Spain, Trinidad.

Watson, J. (1979) *The West Indian Heritage: a History of the West Indies*. John Murray, London.

Williams, E.E. (1964) *History of the People of Trinidad and Tobago*. Praeger, Westport, Connecticut.

Zelinsky, W. (1992) *The Cultural Geography of the United States*. Prentice-Hall, Upper Saddle River, New Jersey.

4 Small-scale Events as a Tool for Sustainability: Prospects and Constraints

Uswathul Hasana* and Sampada Kumar Swain
Pondicherry University, Puducherry, India

Introduction

Event tourism is not a new phenomenon in terms of research and as an industry practice. While not all events are aimed towards tourism development (Getz, 2008), many events are organized and promoted across the globe to boost tourism growth and destination image (Herstein and Berger, 2014). However, their impacts are intense and not always positive. This is especially true in cases of mega-events that have a better reach requiring higher cost at the expense of the host destination population and its environment (Müller, 2015). The lack of an all-inclusive paradigm of multiple stakeholders' interests leads to the lack of consensus on the event outcomes, particularly in mega-events with different stakeholder groups, making them unsustainable (Getz, 2009).

The policy frameworks of such events are designed to favour the tourism and economic perspectives neglecting the sociocultural and environmental dimensions. These forms of neo-liberal governance alienate communities from the events and affect their support needed for long-term success (Higgins-Desbiolles, 2018). Owing to these shortcomings, the sustainability of such events came under crucial criticisms. Studies on events are also criticized for marginalizing their drawbacks regarding social exclusion and economic inequality (Rojek, 2014). On the other hand, the promised tourism development post these mega-events are not always the reality. For example, tourist flows to a country hosting a mega sporting event are exclusively driven by the motivation to watch the event. As a result, the tourist flow would get back to the older phase in the aftermath of the event (Bondarik *et al.*, 2020). In most cases, these mega-events, especially mega sporting events, are unlikely to be held in the same country in the near future, thus putting the expenditure spent on the infrastructure to remain unused.

To overcome the negative impacts caused by mega-events, 'greening' of mega-events was suggested to yield sustainable outcomes (Getz, 2017). However, the 'greening' practices often fail to produce the anticipated results. The primary reasons could be attributed to the fact that these mega-events have far more prioritized objectives set by the sponsors, host countries and other stakeholders (Death, 2011; Dodds and Graci, 2012). The growing popularity of events and concerns regarding their consequences make event tourism one of the growing segments in academia, with more studies on the relationship between events and sustainability (Hall, 2012; Liu *et al.*, 2018). Despite the criticisms, the demand for sustainability is expected to grow in the event sector (Pelham, 2011). With evidence from the impacts of mega-events,

*Corresponding author: uswathulhasana2512@gmail.com

DOI: 10.1079/9781789248685.0004

it is difficult not to question the sustainability of such mega-events. Hence scholars have recommended an alternate form of events conducted on a small scale and with local community representation as more sustainable (Hall, 2012).

As opposed to mega-events, small-scale events involve the participation of a comparatively smaller number of people and require only minimal infrastructure. Small-scale events comprise regional festivals, cultural events, and sports events that happen on a small scale (Lukić *et al.*, 2020). Recent studies in the event tourism literature discuss the potential of small-scale events as a tool towards sustainable development in event tourism. Small-scale sports and cultural events with substantial local representation are gaining importance as a promising sector in event tourism owing to the positive impacts on the host destination and its residents. However, the extent to which these small-scale events can act as a sustainable alternative to mega-events is still not vividly addressed. With the growing consciousness on sustainability and the tourism potential of small-scale events, this chapter aims to study whether event tourism resulting from events that are organized on a small scale can be sustainable or not. It intends to examine the opportunities and limitations of small-scale events in achieving economic, social and environmental sustainability by undertaking a systematic review of the literature. It also discusses about how far such small-scale events can influence the growth of the tourism sector in comparison to mega-events, especially if such small-scale events are to be considered an alternative for mega-events.

Methodology

This research study is based on the systematic review of published literature on event tourism. Research papers to be reviewed were collected using the combination of keywords including 'small-scale events', 'event tourism', 'sustainability' and 'sustainable events' from two search engines (i.e. 'Scopus and Google Scholar'). The research papers collected were analysed using content analysis. The content analysis method is employed to systematically analyse any form of recorded communications (Kolbe and Burnett, 1991).

The chapter is divided into four sections based on the analysis. The first section deals with the relation between event tourism and its effect on 'three pillars of sustainability' (economic, social and environmental). The prospects of small-scale events in contributing to sustainable outcomes are briefly discussed in the second section. The third section elaborates on the challenges and limitations persistent in small-scale events. The last section comprises discussions on the potential of small-scale events as a sustainable event tourism phenomenon.

Event Tourism and the Three Pillars of Sustainability

Despite the fact that events and consequent event tourism have positive impacts on the destination and its communities, the sustainability of such impacts has received mixed reviews. The impacts are criticized for being short-lived and not equally distributed which is contrary to the characteristics of sustainable development (Smith, 2009). Since sustainability is a broad concept, literature dealing with terms covers various dimensions of it. Event tourism literature on sustainability focuses mainly on the economic, social and environmental dimensions. These three dimensions are referred to as the three pillars of sustainability (Purvis *et al.*, 2019).

Event tourism is deemed sustainable if the events that are intended to attract tourists balance their sociocultural, economic and environmental outcomes with the additional requirements to facilitate tourists (Getz, 2009). Previous studies have recommended that compared with mega-events, small-scale events contribute more towards sustainable tourism (Perić *et al.*, 2016). This argument is substantiated by evidence of sustainable social, economic and environmental outcomes of such small-scale events (Gibson *et al.*, 2012; Perić *et al.*, 2016) and criticisms of the consequences of mega-events. Before going further on the prospects and limitations of small-scale events, it is necessary to weigh the impacts of both mega- and small-scale events on sustainability. This section discusses the three pillars of sustainability based on the event outcomes with reference to small-scale and mega-events.

Social sustainability

Sustainability issues in social outcomes are mainly due to the large-scale nature of the events that demand major physical changes and disrupt local life. At the same time, their smaller equivalents need fewer modifications in terms of infrastructure and space (Smith, 2009). Dissimilar to the economic and environmental aspects, the social factors of event outcomes do not entirely depend on the time or aftermath of the events. The process of event conceiving has an equally significant part in influencing social effects that stimulate optimism (Anderson and Holden, 2008). Liu *et al.* (2018) have recommended 20 indicators from a multi-stakeholder perspective to evaluate festival tourism's sustainability, with sociocultural dimensions occupying the top five places. The study puts forward tourists as an important stakeholder who plays an essential role in deciding an event's sustainability. The part of tourists on event sustainability is in turn fuelled by their knowledge and appreciation of the local culture, primarily through the interaction with another vital stakeholder, the local community (Liu *et al.*, 2018).

The sustainable outcomes from small-scale events are mainly contributed through the usage of existing facilities/infrastructure and appropriate local community representation in terms of participation and distribution of benefits. Festival tourists showed better consciousness on spending on local products, which is observed to be encouraged by their better knowledge about local culture inculcated through the participation in local cultural festivals (Nurse, 2001). A study on the impacts of small-scale sports events shows that such events stimulate community pride and get identified due to their authenticity and proximity to the local culture; in contrast, mega sporting events are highly commercialized and globalized (Malchrowicz-Mośko and Poczta, 2018). In addition, small-scale events provide a platform for creative tourism activities that encourage increased social connectedness and higher interactions (Bakas *et al.*, 2019).

Economic sustainability

Economic outcomes in terms of income generation do not make an event economically sustainable (O'Sullivan and Jackson, 2002), as the economic benefits should also be equally distributed to be sustainable. Events demand funding to meet the various infrastructural and other requirements. Many countries use taxpayers' money to invest in such conditions to host events, with a promise of revenue generation to the host community. On the one hand, the fact that such mega-events generate more revenue, outweighing the cost, is valid to an extent; however, the benefits are not always accessible to the host communities. On the other hand, though the economic impacts of small-scale events are minimal, they are far more effective in an equal distribution of economic benefits.

Though hosting mega-events contributes to the economic impacts by generating employment to the communities of host destinations to some extent (Jones and Stokes, 2003), the local representation in formulating the policies and objectives that decide the outcomes of those events are minimal (Burbank *et al.*, 2002). Economic benefits from mega-events are relished by elite groups and it is the common people who bear the costs (Getz, 2017). Contrary to the economic impacts of mega-events, small-scale events produce sustainable economic outcomes. Analysis of the economic impacts of tourism through small-scale sports events revealed that the revenue generated has directly contributed to the host communities and also had indirect economic benefits (Duglio and Beltramo, 2017). Long-term economic benefits for local communities from small-scale sports events are possible as the recurring events ensure reliable visitor flow each year (Vargáné Csobán and Serra, 2014). Similarly, a comparative study on the economic impacts of mega sporting events organized by foreign agencies and locally organized small-scale sports events in South Africa argues that the economic boom created due to the latter proved to be more effective in the long run (Giampiccoli *et al.*, 2015). In general, economic sustainability of small-scale events is based mainly on the economic benefits and how they are distributed among the major stakeholders, especially the local communities. However, for mega-events these economic benefits should be larger than the amount spent on organizing such events as mega-events require increased costs (Müller, 2015).

Environmental sustainability

The environmental impacts of event tourism were a less-explored area compared with the economic and social effects. Impacts of events on the environment should be a major area of concern, given the negative impacts brought by the events to the host environment, since both mega- and minor events use local resources, generate waste and contribute to pollution (Yuan, 2013). Emphasis on sustainability in recent years is slowly paving the way for new frameworks and indicators developed to analyse the environmental impacts of events. One such indicator is the ecological footprint, which helps in quantitatively assessing the environmental capacity against resources used (Collins and Cooper, 2017).

A study assessing the environmental impacts of festivals using ecological footprints states that the major contributor of negative impacts is the travel needed to get to and from the event, especially air travel, and increased usage of private transport services (Collins and Cooper, 2017). Mega-events claiming to follow greening and sustainable practices fail to produce the expected environmental outcomes to be sustainable (Dodds and Graci, 2012). In order to put sustainability into practice, events that induce pro-environmental behaviour of attendees through education are crucial in pursuance towards environmental sustainability (Yuan, 2013). Pro-environmental behaviour induced positive environmental outcomes among the participants of sustainability-focused events organized on a small scale by the community. Yet the major drawback of such events is the fact that they are attended by people who are already committed to sustainability (Mair and Laing, 2013). Unlike the other aspects of sustainability, environmental sustainability is a difficult criterion to achieve provided the consumptive nature of the events regardless of their size (Zifkos, 2014).

Prospects of Sustainable Outcomes from Small-scale Events

When discussing sustainable outcomes, studies insist that it should be for the long term. Long-term benefits from events are possible if the event is a recurring phenomenon and visitors' loyalty is gained, ensuring repeated visits or positive word of mouth. Customer loyalty earned for recurring small-scale events through place attachment will develop the destination attachment (Halpenny *et al.*, 2016). Regional festivals and small-scale sports events are mostly organized at a regular interval and as per the local traditions and customs. Festivals can serve as a medium in promoting the image of small or medium-sized towns (Bradley and Hall, 2006). However, arguments arise about the limited number of visitors these events attract when compared with one-off large events.

Lack of awareness about the small-scale events is attributed to the smaller number of visitors. To overcome this issue, McKercher *et al.* (2006) recommend promoting them by bundling all the small-scale events together. Destinations that host and promote small recurring events throughout the year appeal to different market segments and hold different objectives. In this way, the outcomes of the events can be sustained for a long time (Ziakas and Costa, 2011). Referred to as the 'event portfolio' concept, hosting a number of small-scale events also ensures tourist influx throughout the year irrespective of the peak season (Salgado-Barandela *et al.*, 2021).

Festivals traditionally celebrated in a locality or the ones designed to draw visitors hold opportunities to increase the region's tourism growth (Okech, 2011). For example, an assessment of the impact of Caribbean festivals on tourism revealed that the festivals increased the occupancy rates by creating new tourism seasons and prolonging the stay of tourists. The tourists also showed repeated visits and more extended stays, contributing to local economic development (Nurse, 2001). Often, festivals organized at the local level on a small scale have objectives that are beneficial to development at the regional and community levels (Higham and Ritchie, 2001). Events having objectives primarily focusing on the community and regional development can catalyse local development. The cultural significance of the festivals among the community can be witnessed by the local people taking part in organizing the event and their participation in event activities like music, dance and other performances. Representation of the local community at the forefront of the event encourages better dialogues with the visitors. Particularly if the events allow only a small number of visitors and substantial local representation, then the prospects for social connectedness and networks are higher in such small-scale festivals, creating a favourable environment for social capital (Bakas *et al.*, 2019; Ziółkowska, 2020).

The majority of the small-scale events are the types of events which are traditionally organized and hosted in the destination irrespective of their tourism potential (Lukić *et al.*, 2020). Hence these events are authentic to the host destinations and not staged for the purpose of tourism promotions. For example, a small-scale heritage sport event provides the authenticity which is difficult to witness in homogenized mega sporting events (Malchrowicz-Mośko and Poczta, 2018). Countries like India which is home to a number of regional and local festivals (Colledge, 1999) are gaining recognition for their tourism potential (Sharma, 2019). Tourism growth brought by such regional and local festivals will have a significant contribution in sustaining the cultural values and economic growth of the region (Munjal and Jauhari, 2015; Roy, 2017). Tourism fuelled through festivals like Holi, celebrated across many parts of India, can contribute to the economy of religious and cultural tourism provided such festivals are promoted sustainably (Shinde, 2010).

Constraints and Limitations of Small-scale Events

Even though small-scale events are proposed to deliver benefits to the host destination at minimum cost, there are several challenges and constraints, especially when the benefits are expected to be sustainable. Tourism growth stimulated by small-scale events can extinguish the very essence of the events if left unmonitored. Promotional strategies employed to showcase events as tourism resources have far greater impacts on small-scale events when compared to mega-events (Kelly and Fairley, 2018).

Local or regional events that can boost the tourism sector are often not promoted from a tourism perspective due to potential negative impacts. The economic or other benefits expected from the events lead to compromising the authenticity of the events by stakeholders to meet their own agendas. Rojek refers to this process of compromising the authenticity of an event by stakeholders to meet their agendas as 'Event Appropriation' (Rojek, 2014). Hence, the exposure these events receive resulting from tourism growth is argued as a threat to authenticity (Getz and Page, 2016). Many events which start as a small-scale event grow in scale over a period of time. For example, the Hay Festival, which started as a minor literature festival in Wales with 1000 participants, has become one of the major literary events attracting around 100,000 participants each year (Collins and Cooper, 2017). When a small-scale festival receives recognition resulting in increased visitor influx, there is a risk for both the event and the venue to exceed their carrying capacity. Nonetheless, the potential dangers of overcrowding in small-scale events are still a neglected area of research.

Unlike the mega-events with significant exposure globally, resulting in the surge of visitors to the host destinations, their small-scale counterparts need strategically planned promotional activities to attract outside visitors. Promoting events necessitates knowledge and intelligence of the market, which is very limited in the case of small events (Mehmetoglu and Ellingsen, 2005). Event organizations rely on the minimal budget allocated to them with which it is not feasible to afford resources needed to improve the event quality and promotion (Perić *et al.*, 2016). Evidence from recent studies shows that digital and social media platforms can be used for their promotion and marketing at lower costs. However, research on the effectiveness of social media platforms in event promotion is still in its novel stage (Belenioti *et al.*, 2017; Sharma, 2019).

Irrespective of the presence of such platforms for promotions, small-scale events are outshone by mega-events due to the influx of tourism growth brought by the latter type (Getz, 2008). Outside visitors to small-scale events like cultural and local festivals, in many cases, are tourists who happened to visit the destination at the time of the event. Since their primary motivation is not to participate in the event, it is difficult to evaluate the tourism potential of such events (McKercher *et al.*, 2006).

Community forms a major stakeholder in small-scale events assuming various roles from organizing events to sponsoring them (Perić *et al.*, 2016). Involving members outside the community to organize and promote the event from a tourism perspective may result in a situation threatening the community's existing social network and functioning. Additionally, even in small-scale events, communities' opinions on the events conducted at their locale are sometimes overlooked with the absence of community members' involvement in the decision-making process (Higgins-Desbiolles, 2018). Event tourism sectors dominated by private players fail to prioritize the interests of the communities (Dredge and Whitford, 2011). Even in projects involving the local community, enough attention is not paid to the traditional hierarchical structure of the community that decides the inclusion and exclusion of community members (de Jong and Varley, 2018). Moreover, sustainability of events also implies the need for sustainability of those organizations involved in planning and running the events. In cases of events organized by communities, the sustainability of communities as event organizations cannot be guaranteed since communities are prone to changes as they evolve, which might affect the sustainability of the events (Getz and Andersson, 2008).

Sustainable Tourism through Small-scale Events: an Alternative Strategy?

The growing number of studies on the sustainable outcomes of small events and the criticism on the unsustainable approaches of mega-events puts forward an argument – can small-scale events potentially be a sustainable alternative to mega-events? In a broader sense, the sustainable aspects of small-scale events are fuelled by the notion of the absence or lower rate of negative economic, social and environmental impacts. However, this does not imply tourism development since several such small-scale events do not create tourist inflows that cause such impacts. Hence in many cases, the sustainable nature of small-scale events is simply because there is an absence or negligible share of the tourism aspect.

When focusing on combatting sustainability issues caused by event tourism, emphasis is placed on why such events do not demand special or new infrastructure facilities to meet their requirements. Small-scale events, including sports events, regional festivals and cultural events, can be carried out with the existing infrastructure facilities available in the destinations (Getz and Page, 2016). Nevertheless, an event's sustainability cannot be analysed by relying entirely on the resource it demands. The concerns created during and after the event, such as problems in waste management and the rising level of pollution created by increasing numbers of visitors, are still a risk to sustainability irrespective of the size of the event (Yuan, 2013). Also, bidding for mega-events such as the Olympics is done to carry out the development projects of the destinations (Burbank *et al.*, 2002). Venues of such mega-events get the opportunity for regeneration either through funding or the events catalysing the existing development projects (Jones and Stokes, 2003). Therefore, the destinations which need such regeneration programmes opt for bidding on mega-events.

Apart from their sustainability, another vital issue that should be considered is to what extent these small-scale events can attract visitors to generate outcomes. Though it is evident that events conducted on a small scale cannot produce the same results in terms of tourist flow as the mega-events (Herstein and Berger, 2014), a standard visitor influx is necessary to sustain the event. A limited number of tourists to events has negative implications on the economic outcomes owing to less tourist spending. Small-scale events failing to attract tourists face fewer opportunities for income generation, affecting the economic impacts (Egresi and Kara, 2014). Devoid of mega-sponsors, small-scale events rely on limited resources that are inadequate to create exposure. Even though small-scale events with considerable local popularity may never be promoted as a prime attraction to visitors, they can add value to the region's existing destination image, thereby increasing the tourism growth (McKercher *et al.*, 2006; Kaplanidou and Vogt, 2007; Hernández-Mogollón *et al.*, 2018). However, overcrowding, a familiar scenario in mega-events, is the source of adverse outcomes, including congestion, unavailability of space, and inaccessibility of services by locals (Fredline and Faulkner, 1998). Small-scale events are designed to attract a very limited market section of tourists and hence can avoid the impacts created due to enormous visitor influx as occurs in mega-events. Because of the narrow market segment of small-scale/minor events, they are regarded as a suitable strategy in promoting and branding small cities whereas large cities need larger events for their branding or rebranding (Herstein and Berger, 2014).

Another essential aspect that should be considered when putting forth small-scale events as an alternative strategy to achieve sustainable tourism development is the political factors associated with hosting mega-events. The primary reason that many countries compete to host such mega-events is not only tourism growth but the political outcomes. Generally, localized minor events cannot deliver the political appeal desired by the host countries at a global level (Hall, 2012). Therefore, these countries might prefer small-scale events as supplementary sustainable initiatives rather than sustainable alternatives to mega-events. This can be a significant challenge as collective policy reforms on governance are still lacking to address event tourism in general and sustainability issues in particular (Dredge and Whitford, 2011). The policy reforms also need to be exclusive to small-scale events as their objectives and the anticipated outcomes are not similar to mega-events (Kelly and Fairley, 2018).

It would be irrational to generalize small-scale events as a sustainable alternative for mega-events without effective strategies to regulate their growth. Many destinations depend on mega-events to reach or to retain their places in the highly competitive environment. Hence, despite the fact that small-scale events can produce sustainable outcomes, a holistic approach at policy level is still needed to implement them as a sustainable alternative to mega-events (Hall, 2012).

Conclusion

Despite their potential, tourism benefits from small-scale events are yet to be tapped because event producers lack the resources in the form of funding and expertise, which affects the strategic planning and promotion of the events (Higham and Ritchie, 2001). There is a consensus among studies stating that small-scale events can bring the expected tourism growth only if proper promotional strategies are employed (Nurse, 2001; Lukić *et al.*, 2020) and also how combining a series of small-scale events can magnify the influx of tourists (Ziakas and Boukas, 2016). In regard to such promotional and marketing activities, a collective approach from the stakeholders that possess the financial resources, such as government agencies and business owners offering tourism-related services, is recommended (Liu *et al.*, 2018). Strategies with a collaborative approach between the tourism enterprises and event organizers are crucial since the former's primary objective is to bring in more tourists to the events, and the latter focuses more on the cultural or social importance of the events (Kelly and Fairley, 2018).

Though there are a good number of studies on small-scale events, most of these are focused on sports events (Lukić *et al.*, 2020). For better policies to promote small-scale events, there is a need for more empirical studies focusing on the impacts and potential of such events. Event tourism from small-scale events has the potential to contribute to sustainable outcomes. Yet sustainability can be achieved only if these outcomes could be continued in the long run, which is arguable, provided the absence of policies and governance focusing on event tourism.

References

Anderson, B. and Holden, A. (2008) Affective urbanism and the event of hope. *Space and Culture* 11(2), 142–159. DOI: 10.1177/1206331208315934.

Bakas, F.E., Duxbury, N., Remoaldo, P.C. and Matos, O. (2019) The social utility of small-scale art festivals with creative tourism in Portugal. *International Journal of Event and Festival Management* 10(3), 248–266. DOI: 10.1108/IJEFM-02-2019-0009.

Belenioti, Z.C., Gkarane, S. and Vassiliadis, C. (2017) The contribution of social media to small scale events: boosting the city branding and tourism demand through Thessaloniki Food Festival. Paper presented at the 5th International Conference on Contemporary Marketing Issues ICCMI (International Conference on Contemporary Marketing Issues), 21–23 June, Thessaloniki, Greece.

Bondarik, R., Pilatti, L.A. and Horst, D.J. (2020) Sports mega-events and overestimated promises: the case of the 2014 FIFA World Cup in Brazil. *Journal of Tourism and Cultural Change* 18(3), 351–367. DOI: 10.1080/14766825.2020.1727488.

Bradley, A. and Hall, T. (2006) The festival phenomenon: festivals, events and the promotion of small urban areas. In: Bell, D. and Jayne, M. (eds) *Small Cities: Urban Experience Beyond the Metropolis*. Routledge, Abingdon, UK, pp. 77–90.

Burbank, M.J., Andranovich, G. and Heying, C.H. (2002) Mega-events, urban development, and public policy. *The Review of Policy Research* 19(3), 179–202. DOI: 10.1111/j.1541-1338.2002.tb00301.x.

Colledge, R. (1999) *Mastering World Religions*. Macmillan Education, London. DOI: 10.1007/978-1-349-14329-0.

Collins, A. and Cooper, C. (2017) Measuring and managing the environmental impact of festivals: the contribution of the Ecological Footprint. *Journal of Sustainable Tourism* 25(1), 148–162. DOI: 10.1080/09669582.2016.1189922.

de Jong, A. and Varley, P. (2018) Food tourism and events as tools for social sustainability? *Journal of Place Management and Development* 11(3), 277–295. DOI: 10.1108/JPMD-06-2017-0048.

Death, C. (2011) 'Greening' the 2010 FIFA World Cup: environmental sustainability and the mega-event in South Africa. *Journal of Environmental Policy & Planning* 13(2), 99–117. DOI: 10.1080/1523908X.2011.572656.

Dodds, R. and Graci, S. (2012) Greening of the Pride Toronto Festival: lessons learned. *Tourism Culture & Communication* 12(1), 29–38. DOI: 10.3727/109830412X13542041184739.

Dredge, D. and Whitford, M. (2011) Event tourism governance and the public sphere. *Journal of Sustainable Tourism* 19(4–5), 479–499. DOI: 10.1080/09669582.2011.573074.

Duglio, S. and Beltramo, R. (2017) Estimating the economic impacts of a small-scale sport tourism event: the case of the Italo-Swiss Mountain Trail CollonTrek. *Sustainability* 9(3), 343. DOI: 10.3390/su9030343.

Egresi, I. and Kara, F. (2014) Economic and tourism impact of small events: the case of small-scale festivals in Istanbul, Turkey. *Studia Ubb Geographia* LIX(1), 47–64.

Fredline, E. and Faulkner, B. (1998) Resident reactions to a major tourist event: the Gold Coast Indy Car Race. *Festival Management and Event Tourism* 5(4), 185–205. DOI: 10.3727/106527098791784475.

Getz, D. (2008) Event tourism: definition, evolution, and research. *Tourism Management* 29(3), 403–428. DOI: 10.1016/j.tourman.2007.07.017.

Getz, D. (2009) Policy for sustainable and responsible festivals and events: institutionalization of a new paradigm. *Journal of Policy Research in Tourism, Leisure and Events* 1(1), 61–78. DOI: 10.1080/19407960802703524.

Getz, D. (2017) Developing a framework for sustainable event cities. *Event Management* 21(5), 575–591. DOI: 10.3727/152599517X15053272359031.

Getz, D. and Andersson, T.D. (2008) Sustainable festivals: on becoming an institution. *Event Management* 12(1), 1–17. DOI: 10.3727/152599509787992625.

Getz, D. and Page, S.J. (2016) Progress and prospects for event tourism research. *Tourism Management* 52, 593–631. DOI: 10.1016/j.tourman.2015.03.007.

Giampiccoli, A., Lee, S.S. and Nauright, J. (2015) Destination South Africa: comparing global sports mega-events and recurring localised sports events in South Africa for tourism and economic development. *Current Issues in Tourism* 18(3), 229–248. DOI: 10.1080/13683500.2013.787050.

Gibson, H.J., Kaplanidou, K. and Kang, S.J. (2012) Small-scale event sport tourism: a case study in sustainable tourism. *Sport Management Review* 15(2), 160–170. DOI: 10.1016/j.smr.2011.08.013.

Hall, C.M. (2012) Sustainable mega-events: beyond the myth of balanced approaches to mega-event sustainability. *Event Management* 16(2), 119–131. DOI: 10.3727/152599512X13343565268294.
Halpenny, E.A., Kulczycki, C. and Moghimehfar, F. (2016) Factors effecting destination and event loyalty: examining the sustainability of a recurrent small-scale running event at Banff National Park. *Journal of Sport & Tourism* 20(3–4), 233–262. DOI: 10.1080/14775085.2016.1218787.
Hernández-Mogollón, J.M., Duarte, P.A. and Folgado-Fernández, J.A. (2018) The contribution of cultural events to the formation of the cognitive and affective images of a tourist destination. *Journal of Destination Marketing & Management* 8, 170–178. DOI: 10.1016/j.jdmm.2017.03.004.
Herstein, R. and Berger, R. (2014) Cities for sale: how cities can attract tourists by creating events. *The Marketing Review* 14(2), 131–144. DOI: 10.1362/146934714X14024778816832.
Higgins-Desbiolles, F. (2018) Event tourism and event imposition: a critical case study from Kangaroo Island, South Australia. *Tourism Management* 64, 73–86. DOI: 10.1016/j.tourman.2017.08.002.
Higham, J.E.S. and Ritchie, B. (2001) The evolution of festivals and other events in rural southern New Zealand. *Event Management* 7(1), 39–49. DOI: 10.3727/152599501108751461.
Jones, M. and Stokes, T. (2003) The Commonwealth Games and urban regeneration: an investigation into training initiatives and partnerships and their effects on disadvantaged groups in East Manchester. *Managing Leisure* 8(4), 198–211. DOI: 10.1080/1360671032000148576.
Kaplanidou, K. and Vogt, C. (2007) The interrelationship between sport event and destination image and sport tourists' behaviours. *Journal of Sport & Tourism* 12(3–4), 183–206. DOI: 10.1080/14775080701736932.
Kelly, D.M. and Fairley, S. (2018) What about the event? How do tourism leveraging strategies affect small-scale events? *Tourism Management* 64, 335–345. DOI: 10.1016/j.tourman.2017.09.009.
Kolbe, R.H. and Burnett, M.S. (1991) Content-analysis research: an examination of applications with directives for improving research reliability and objectivity. *Journal of Consumer Research* 18(2), 243. DOI: 10.1086/209256.
Liu, C.-R., Lin, W.-R., Wang, Y.-C. and Chen, S.-P. (2018) Sustainability indicators for festival tourism: a multi-stakeholder perspective. *Journal of Quality Assurance in Hospitality & Tourism* 20(3), 296–316. DOI: 10.1080/1528008X.2018.1530165.
Lukić, D., Stanković, S., Petrović, M.D., Radovanović, M.M., Vuković, D. *et al.* (2020) How can small-scale events contribute to the tourism progress of an undeveloped border area? Lesson from Eastern Serbia. *GeoJournal of Tourism and Geosites* 30(2 supplement), 905–912. DOI: 10.30892/gtg.302spl17-521.
Mair, J. and Laing, J.H. (2013) Encouraging pro-environmental behaviour: the role of sustainability-focused events. *Journal of Sustainable Tourism* 21(8), 1113–1128. DOI: 10.1080/09669582.2012.756494.
Malchrowicz-Mośko, E. and Poczta, J. (2018) A small-scale event and a big impact—is this relationship possible in the world of sport? The meaning of heritage sporting events for sustainable development of tourism—experiences from Poland. *Sustainability* 10(11), 4289. DOI: 10.3390/su10114289.
McKercher, B., Mei, W.S. and Tse, T.S.M. (2006) Are short duration cultural festivals tourist attractions? *Journal of Sustainable Tourism* 14(1), 55–66. DOI: 10.1080/09669580608668591.
Mehmetoglu, M. and Ellingsen, K.A. (2005) Do small-scale festivals adopt 'market orientation' as a management philosophy? *Event Management* 9(3), 119–132. DOI: 10.3727/152599505774791176.
Müller, M. (2015) What makes an event a mega-event? Definitions and sizes. *Leisure Studies* 34(6), 627–642. DOI: 10.1080/02614367.2014.993333.
Munjal, S. and Jauhari, V. (2015) Leveraging the cultural and economic value of fairs and festivals in India. *Worldwide Hospitality and Tourism Themes* 7(4), 417–423. DOI: 10.1108/WHATT-03-2015-0006.
Nurse, K. (2001) *Festival Tourism in the Caribbean: an Economic Impact Assessment*. Inter-American Development Bank, Washington, DC.
Okech, R.N. (2011) Promoting sustainable festival events tourism: a case study of Lamu Kenya. *Worldwide Hospitality and Tourism Themes* 3(3), 193–202. DOI: 10.1108/17554211111142158.
O'Sullivan, D. and Jackson, M.J. (2002) Festival tourism: a contributor to sustainable local economic development? *Journal of Sustainable Tourism* 10(4), 325–342. DOI: 10.1080/09669580208667171.
Pelham, F. (2011) Will sustainability change the business model of the event industry? *Worldwide Hospitality and Tourism Themes* 3(3), 187–192. DOI: 10.1108/17554211111142149.
Perić, M., Đurkin, J. and Wise, N. (2016) Leveraging small-scale sport events: challenges of organising, delivering and managing sustainable outcomes in rural communities, the case of Gorski kotar, Croatia. *Sustainability* 8(12), 1337. DOI: 10.3390/su8121337.

Purvis, B., Mao, Y. and Robinson, D. (2019) Three pillars of sustainability: in search of conceptual origins. *Sustainability Science* 14(3), 681–695. DOI: 10.1007/s11625-018-0627-5.

Rojek, C. (2014) Global event management: a critique. *Leisure Studies* 33(1), 32–47. DOI: 10.1080/02614367.2012.716077.

Roy, S. (2017) Community based events and tourism promotion: a study with special reference to Jonbeel Mela in Assam. *International Journal of Management Studies* IV(1), 66–74.

Salgado-Barandela, J., Barajas, A. and Sanchez-Fernandez, P. (2021) Sports event portfolios: an innovative tool and a new management paradigm. *International Journal of Sports Marketing and Sponsorship* (ahead-of-print). DOI: 10.1108/IJSMS-02-2021-0024.

Sharma, A. (2019) Festivals for sustainable tourism development: a case study of Hadoti region, Rajasthan. In: Mair, J. (ed.) *The Routledge Handbook of Festivals*. Routledge, Abingdon, UK, pp. 366–373.

Shinde, K.A. (2010) Managing Hindu festivals in pilgrimage sites: emerging trends, opportunities, and challenges. *Event Management* 14(1), 53–67. DOI: 10.3727/152599510X12724735767598.

Smith, A. (2009) Theorising the relationship between major sport events and social sustainability. *Journal of Sport & Tourism* 14(2–3), 109–120. DOI: 10.1080/14775080902965033.

Vargáné Csobán, K. and Serra, G. (2014) The role of small-scale sports events in developing sustainable sport tourism – a case study of fencing. *Applied Studies in Agribusiness and Commerce* 8(4), 17–22. DOI: 10.19041/APSTRACT/2014/4/3.

Yuan, Y.Y. (2013) Adding environmental sustainability to the management of event tourism. *International Journal of Culture, Tourism and Hospitality Research* 7(2), 175–183. DOI: 10.1108/IJCTHR-04-2013-0024.

Ziakas, V. and Boukas, N. (2016) The emergence of "small-scale" sport events in "small island" developing states: towards creating sustainable outcomes for island communities. *Event Management* 20(4), 537–563. DOI: 10.3727/152599516X14745497664479.

Ziakas, V. and Costa, C.A. (2011) The use of an event portfolio in regional community and tourism development: creating synergy between sport and cultural events. *Journal of Sport & Tourism* 16(2), 149–175. DOI: 10.1080/14775085.2011.568091.

Zifkos, G. (2014) Sustainability everywhere: problematising the "sustainable festival" phenomenon. *Tourism Planning & Development* 12(1), 6–19. DOI: 10.1080/21568316.2014.960600.

Ziółkowska, J. (2020) Enhancing social capital development throughout cultural events. A case study of Eastern Poland. *GeoJournal of Tourism and Geosites* 29(2), 450–459. DOI: 10.30892/gtg.29206-481.

5 Festivals in Terms of Cultural Sustainability: a Study from the Perspective of Tourist Guides

Funda Ön[1]*, Hande Akyurt Kurnaz[2] and Nisan Yozukmaz[3]
[1]Mugla Sıtkı Kocman University, Mugla, Turkey; [2]Bolu Abant İzzet Baysal University, Bolu, Turkey; [3]Pamukkale University, Denizli, Turkey

Introduction

The demand for tourism activities in the world has increased significantly over the last 60 years (UNWTO, 2015). Meanwhile, globalization in the world of travel and tourism as well as the development of technology and the increase in disposable income have intensified competition between tourism destinations (Mariani and Baggio, 2012). In this context, destinations take effective steps by offering interesting and memorable tourist experiences (Snepenger *et al.*, 2006; Coghlan *et al.*, 2012) in order to survive within this competitive industry. As the competition increases, activities and accordingly academic research on sustainability in tourism increase too.

Along with the concept of sustainability, it has become possible to ensure continuity in tourism, protect resources, keep cultural values alive and increase the efforts towards environmental protection. The importance of tourist guides is substantial in terms of retaining and conveying cultural values. Thanks to tourist guides, cultural values are conveyed to different tourist groups and in this way a contribution to cultural sustainability is achieved. Culture is one of the most important legacies left to future generations. In this respect, there are many factors within the concept of cultural sustainability. One of these factors is festivals, which have an important share in tourism.

In this chapter the concept of cultural sustainability, which is a part of the notion of sustainability in tourism, is discussed and festivals are evaluated in terms of cultural sustainability. In this evaluation, opinions of the tourist guides working in seven different geographical regions of Turkey have been analysed.

Cultural Sustainability

Before explaining the concept of cultural sustainability, it is helpful to start with the concept of culture. There are many definitions of culture. The main reason for this is that culture is a notion that develops with societies. Culture refers to the arts and other human intellectual achievement and the customs and civilization of a particular time or people. Humans and culture are two inseparable parts that are fed from each other. At the beginning of many articles on culture, the difficulty of defining the concept of

*Corresponding author: fundaon@gmail.com

 Festival and Event Tourism: Building Resilience and Promoting Sustainability (eds A. Sharma *et al.*)
DOI: 10.1079/9781789248685.0005

culture and the existence of many definitions are discussed (Akın, 2007). However, the difficulty of making a precise definition of culture is not primarily due to the lack of existing knowledge and materials, but to the fact that this concept has a wide range and scope (Eroğlu, 2006). The term culture is derived from the Latin word '*colere*' which means order or maintenance. The word '*cultura*' was first used by Voltaire in the meaning of formation, development and cultivation of the human mind (Kartarı, 2006). The concept of culture was first used by Edward Taylor in his book *Primitive Culture* in 1871 (Burke, 2006). According to Taylor, culture is 'a complex ensemble including the knowledge, art, ethics, traditions and other similar skills and habits acquired by human beings as a member of society' (Güvenç, 1985). There are two basic elements for the actualization of the concept of cultural sustainability. The first element is establishing the required policies and the second is people's embracing their culture (Konuk, 2009).

Cultural sustainability is the interaction of cultural values from past eras with other cultures in order to renew themselves and be transmitted to new generations (Cebeci and Çakılcıoğlu, 2002). Cultural sustainability has a double effect. Its first impact is that the ecological, economic and social issues of sustainability can only be understood within a culture horizon covering opinions, images and concepts that are used for defining sustainability. Its second impact is that sustainability is an issue in itself, because there are traditions, perception types, cognition methods and conceptualization of knowledge related to it (Mignolo, 2011). Cultures depend on memory-supported modes of transmission; they exist in a continuous process of transformation through reclaiming, reworking, reassessing, revitalizing and reconstructing a group's collected and collective heritage. But this also means that the future of cultural memory and heritage is always precarious. Thus, cultural memory can easily be neglected and destroyed (Assman, 2001).

Considering that culture reflects life, it is very important to ensure its sustainability and transfer it from past to present (Konuk, 2009). In order to preserve cultural heritage and transfer it to the next generations in cultural sustainability, it is necessary to spread heritage values, for cultural heritage to be recorded with the required research and archiving studies, and to ensure the sustainability of tourism (Pelit *et al.*, 2018). The concept of sustainability emerges as a process. This concept not only relates to physical items (e.g. historical artefacts), it also relates to the social and cultural structure (e.g. traditions and customs). So regarding the survival and maintenance of cultural values, cultural sustainability includes all factors that have cultural value to a society such as historical and cultural artefacts, traditions and customs, festivals, events, etc.

Festivals and Festival Tourism within the Scope of Cultural Sustainability

According to Kucharska and Kowalczyk (2018), tourism development in a particular destination undergoes many changes as a result of tourist travel and trips. These changes are related to spatial development, economy, sociocultural space, political-administrative institutions and the natural environment. Tourism can have positive or negative impacts on the environment. Its positive impacts for people include functions such as relaxation, novelty or employment opportunities. Transferring culture to future generations is one of the important positive functions of tourism (Getz, 2010). At this point, the concept of cultural sustainability emerges. Tourism has a great impact within the scope of cultural sustainability, which is defined as transferring culture to the future. One of the most effective components of this concept is festivals.

Festivals have a positive impact on cultural heritage, which is another element of tourist attractions. They are produced as a part of human culture. They consist of all tangible and intangible factors as well as being attractions themselves in a specific destination. The festivals are mostly organized in places such as museums, galleries, theatres, music halls or places of intangible cultural heritage on which they have a positive effect too (Cudny, 2013).

Festivals are an important source of motivation for tourism and are prominently included in the development and marketing plans of many destinations. From a different perspective, festivals have become a rapidly growing market in the tourism industry. Festival tourism emerges as a result of visitors' participation in these planned events and festivals. Festivals not only promote tourism, but also play an important role in urban regeneration, cultural development, encouraging national identities, and ensuring cultural sustainability (Getz, 2008).

Festivals are a fast-developing phenomenon and play a major role in tourism development of regions. In this context, research on festivals has become very popular recently. As mentioned above, attending a festival can be considered as a separate type of tourism, called festival tourism, which can also be discussed under the concept of special interest tourism (Cudny *et al.*, 2012). Festivals have certain features that make them art and culture-related events reflecting significant aspects about the lives of societies such as being diverse and unique. Sometimes contests and competitions can also take place within the scope of festivals (Cudny, 2013).

Research on festival tourism started to appear in tourism-related journals in the late 1980s and early 1990s (Laing, 2018). Festival tourism has become both a field of study and a globally important sector of economy (Getz, 2012). At the destination level, festival tourism refers to the development and marketing of planned events (Getz, 2013; Laing, 2018). Since the first half of the 2000s, festival tourism has become a multifaceted concept aiming to reveal all aspects of a destination (Larson, 2009). Not only is festival tourism regarded as a strategic tourism area, but it also becomes a factor that differentiates a destination from its competitors (Getz and Page, 2016) such as Glastonbury Festival held near the town of Glastonbury in Somerset, UK.

Festival tourism can also have negative effects on cultural heritage in terms of its intangible components. Festivals in particular should continue to reflect and preserve cultural heritage without losing the originality and uniqueness of that heritage. Modified or misrepresented cultural heritage causes visitors to have misleading information about destinations (Getz, 2010). If festivals are trivialized or adapted to the tastes of mass audiences, they move away from reality and cause a misunderstanding of the local culture. Festival tourism may also lead to a faster deterioration of some factors such as transportation infrastructure as a result of heavy visitor traffic. In order to prevent this situation, careful and supervised event planning and management should take place (Cudny, 2013).

The Method of the Study

Aiming to examine the festivals in Turkey within the framework of cultural sustainability from the tourist guides' perspective, this study adopted a qualitative approach. The semi-structured interview technique was used as a data collection tool due to the convenience of analysis, opportunity given to participants to express themselves freely, and to collect in-depth information about them (Büyüköztürk *et al.*, 2016, p. 154). This technique is used frequently in exploratory research and is the type of interview in which the outline is predetermined, questions are not sequenced and during which new questions can be produced or some questions can be removed (Yüksel and Yüksel, 2004). Accordingly, in this study the questions in the semi-structured interview form were determined in accordance with the purpose. During the preparation process of the interview form, relevant literature was reviewed. The final version of the interview form was applied after it had been examined by two different researchers working in the relevant field. The interview form included questions about festivals, cultural sustainability and efforts made towards maintaining cultural sustainability.

This research was based on the opinions and ideas of tourist guides working in Turkey. For data collection, tourist guides were interviewed. The reasons for choosing tourist guides for the study sample were: (i) their knowledge about destinations; (ii) their active participation in festivals organized in these destinations; and (iii) their being informed about the current situation of these destinations. The interviews were conducted on a voluntary basis between 11 November and 20 December 2020. Convenience sampling was used so that the data collection process could be quicker, more practical and objective. In total, seven tourist guides participated in the research, each of whom was from a different region of Turkey. The interviews were conducted by telephone due to the pandemic and the distance between researchers and tourist guides. The interviews lasted 20 minutes on average and were recorded. These recordings were then transcribed by the researchers.

The data obtained from the interviews were analysed by descriptive-interpretative analysis in which the data were summarized in relation to predetermined themes. Also, direct quotations were used in the findings section of this chapter as they could be frequently used to convey the opinions of the participants directly (Yıldırım and Şimşek, 2013). The basic principle was to represent what was researched accurately rather than presenting all of the obtained data (Ekiz, 2015, p. 76). During the descriptive analysis, codes were allocated to different themes; the data assigned to themes were described with a comprehensible narration and supported by direct quotations. During the interpretative analysis, appropriate evaluations were made in accordance with the statements of the participants.

Findings

After the analysis of the interviews with tourist guides, the findings were gathered under three main themes. These were 'festivals', 'cultural sustainability' and 'efforts towards cultural sustainability'.

Festivals

During the interviews, the interviewees were first asked which festivals they attended to fulfil the aim of determining their participation level. Only one of the interviewees claimed that she had not attended any festival. The festivals that the interviewees had attended were as follows:

- Shab-i Arus/Seb-i Arus;
- International Konya Mystic Music Festival;
- Beysehir Culture, Art, Tourism and Lake Festival;
- International Nasreddin Hoca Festivities;
- Datça Almond Blossom Festival;
- Gokova under my pedals Bicycle Tour;
- Karia Shoreline Bicycle Tour;
- Gokova Bicycle Tour;
- Gökceada Local Tastes Festival;
- Bozcaada Vintage Festival;
- Adana Orange Blossom Festival;
- Aspendos Opera and Ballet Festival;
- Inlice Music Festival;
- Uzumlu Mushroom Festival; and
- International Zurnazen Festival.

The interviewees asserted that the festivals would affect the natural and cultural environment of the regions in either a positive or negative way. The opinions of the interviewees are presented in Table 5.1.

Cultural sustainability

Within the scope of the study, it was aimed to determine the perceptions of the participants regarding the concept of cultural sustainability. For this aim, the participants were primarily asked about the first words or concepts regarding cultural sustainability that came to their mind. As a result of the interviews, the perceptions of the participants regarding this concept were gathered under themes such as: (i) customs and traditions; (ii) folk dances; (iii) local foods; (iv) handicrafts; (v) continuity; (vi) authenticity; (vii) promotion; (viii) the relationship between people and the environment; (ix) strategy; (x) human health and comfort; (xi) architecture; and (xii) politics. Some interviewees expressed their thoughts on the meaning of cultural sustainability as follows:

> 'The restoration works to ensure the longevity of historical immovable buildings; ensuring the continuity of traditions and customs by preserving the pure structure; conveying food culture and gastronomy to next generations; keeping and preserving original characteristics of ancient resources for [the] future without changing their meaning and protecting verbal culture and accent structure ...' (I1)

> 'Objective education starting from the earliest stages; enhancing the reputation of verbal academic fields with government support if necessary; preventing alienation by suitable designs for [the] environment; taking measures against overpopulation in metropolitan areas ...' (I2)

> 'Protecting intangible cultural values; taking measures to prevent economic and technological development from giving harm to culture; preserving cultural diversity; trying to recover lost cultural values;

Table 5.1. Interviewees' comments on the effects of festivals on the natural and cultural environment of destinations.

I[a]	Natural environment	Cultural environment
I1	Festivals contribute to the development of the physical environment ...	Thanks to the festivals, local people become more aware of the cultural values of their region, engage in a deeper research about it and make more effective promotions of the region in order to reflect their own local culture to the festival visitors.
I2	I observe that festivals can damage the natural environment because they cause a very intense flux of tourists.	I believe that a kind of artificial cultural atmosphere is produced for touristic purposes.
I3	It depends on the type of the festival. For example, the high number of boats participating in a yacht festival leads to pollution of the sea; or a youth festival can cause environmental pollution due to the accumulation of tons of garbage in the camping area.	During the festival, different cultural interactions between locals and visitors or among visitors can occur in the region.
I4	I think festivals have a good impact in terms of preservation and development of a region, but a diverse impact in terms of popularization. It is a fact that festivals have negative impacts as well as positive ones in regions where the natural environment is a source of income. But you can clearly see the negative impacts, especially with the increase in the tourism demand and thus in the housing demand due to the festivals held in the Black Sea region.	Festivals provide mobilities in tourism and so this can change the socio-economic structure of a region and may cause cultural changes. In addition to this, it can be said that festivals are very beneficial in the preservation of culture.
I5	Festivals can affect [the natural environment] unless the participants care about it.	The interaction of travellers and locals during festivals creates an absolute cultural exchange.
I6	[The natural environment] may be affected in a different way. For example, Local Tastes Festival may cause the promotion and marketing of the region and an increase in population. Eventually, the destination could start to grow and the natural environment could be negatively affected.	I think rather than affecting the cultural environment, festivals protect it. Recently emerged festivals can provide cultural diversity, of course.
I7	They mostly affect [the natural environment] positively. They increase sensitivity in landscaping, environmental concerns and have an impact for sure.	Festivals can lead to the development of local people in terms of their skills such as making different presentations to visitors or improving their language skills to appeal to foreign tourists.

[a]I refers to Interviewee.

explaining cultural richness to the new generation in a way that will attract their attention.' (I4)

One interviewee explained cultural sustainability with examples of cultural attractions of particular destinations, as follows:

'Like the dastar weaving in the town Uzumlu, highland tourism in the Black Sea Region, local markets in tourism destinations such as Fethiye Friday Local Market, food festivals, foodie tours to Antep, year-round events or travels of people from different religions.' (I7)

The participants thought that festivals played an important role in ensuring cultural sustainability. Some of their opinions are as follows:

'the main purpose of festivals is to protect up-to-datedness of original culture by spreading it to wider audience.' (I1)

'In terms of continuity of sustainability, people from different cultures are provided with the opportunity to promote their own culture through festivals.' (I3)

'[Festivals] allow a specific culture to appeal to a wider audience.' (I4)

'Festivals contribute to cultural sustainability as long as their authenticity and originality can be preserved and they are authentic and unique.' (I5)

'[As the festival] is a tradition that goes back hundreds of years, of course, it has [a role in cultural sustainability]. At least, it ensures intergenerational interaction.' (I7)

Considering the impact of festivals on cultural sustainability, most of the participants stated that they had an impact and said:

'With these events, many cultural elements that are about to be forgotten are transferred to the next generations.' (I3)

'Whatever the subject of cultural sustainability is, it undoubtedly has an active impact in terms of the recognition of and perceptions towards its subject.' (I5)

The participating tourist guides explained the importance of festivals in terms of cultural sustainability under other themes such as: (i) festivals as a tool for promotion; (ii) cultural interaction; and (iii) protection and recovery of values. Some participants talked about the importance of cultural sustainability as follows:

'(The festival) is a tool for promoting culture and keeping it up-to-date.' (I1)

'The main point is that people come together in different regions of a country and in different parts of the world and share their culture in interaction.' (I3)

'Yes, (the festival) is very important. If you examine the festivals held in certain locations, you will see that they have become a natural protection method in order to revive the values almost extinct and to protect various natural features and geography.' (I4)

'Festivals also carry the values of the past to [the] present and of course the future. From this point of view, the protection of present values is important as they convey information to people who are foreign to that region. It is necessary to raise awareness of conservation in order to ensure cultural sustainability, to establish conservation policies and to recover the lost handicrafts by transferring them to future generations.' (I6)

'[The festival] can help sustain [the] traditional lifestyle and, for example, keep language dialects, temple habits or [the] past life alive.' (I7)

The efforts towards cultural sustainability

From the interviews with the participant tourist guides, some information was obtained about whether efforts towards cultural sustainability were carried out through festivals organized in the destinations. It was found out that some efforts had been made, albeit partially, but were insufficient. Some of the participants' opinions are as follows:

'In this region, local authorities and municipalities offer moral and material support. Including the smallest area, events and festivals are organized annually all over the province and continuity of these promotions are considered important.' (I1)

'No, [there is not any effort] as far as I have observed. I am convinced that they are artificial and only commercial.' (I2)

'In the festivals organized in our country, the cultural environment of a region is highlighted and promoted; thus, it can reach a larger number of people.' (I3)

'It may vary from region to region. In the regions where I work, the present conditions are tried [trying] to be preserved.' (I6)

'I don't think there is enough effort. Present efforts continue as some [voluntary] associations work locally. Of course, it is very insufficient.' (I7)

The suggestions of the tourist guides for what can be done to ensure cultural sustainability are presented in Table 5.2.

Conclusion

Tourism has many positive and negative effects due to its characteristics. It is possible to minimize its negative effects by ensuring sustainability in tourism. Sustainable tourism has become a popular research topic in recent years. In particular, the careful use of resources, reducing environmental pollution, diminishing the damage to cultural and natural values, raising environmental awareness in tourism and spreading tourism activities all year round instead of keeping it seasonal are among the

Table 5.2. Interviewees' suggestions for efforts towards cultural sustainability through festivals.

I[a]	Suggestions
I1	All authorities and associations should work in collaboration in local festivals and be able to appeal to more people [visitors].
I2	It may be useful to limit [festivals or events] to certain activities that would be determined by an expert selection committee.
I3	With more frequent arrangements, making efforts to address our cultural heritage sinking into oblivion and to transfer it to the next generations will yield significant results.
I4	First of all, it should be imposed on local people that visitors are not seen as consumers or customers; diversity of festival contents should be increased; the number of festivals which are suitable for the region considering geographical and climatic differences should be increased.
I5	Efforts towards informing and raising awareness of both visitors and hosts will be useful.
I6	All kinds of attempts to produce and sell local handicrafts or products will affect local people in cultural and economic terms and transferring these values that are specific and unique to the region to future generations will maintain sustainability.
I7	The first thing to do is to inform local people and involve them in all kinds of works. There is a very good proverb in our culture: 'A mill does not rotate with transported water.' Of course, professional support should be provided for commercial productions. Logistics support should be provided for sale of these products. Local authorities should be informed and involved in these efforts.

[a]I refers to Interviewee.

most important sustainable tourism issues. Sustainability is especially important for countries or destinations that derive part of their economic revenue from the tourism industry. With sustainable tourism, infrastructure and superstructure facilities are used more efficiently, efforts are made to minimize environmental degradation, and the effects of large crowds of visitors, in particular high densities of tourists that accumulate in high season, can be diminished.

One of the main resources within sustainable tourism is cultural values that are rare works reflecting the beauty of societies, providing information about civilizations and taking people on a journey in history. In this sense, it is important to protect and preserve cultural values. Cultural sustainability, which means preservation of cultural values and transferring them to the next generations, has a fundamental impact on keeping cultures alive. Festivals are one of the organized events where cultural sustainability can be most effective, transferring culture to future generations. Festivals are events that are organized for different purposes and present attendees with a variety of topics related to entertainment, education, environment and art. Undoubtedly, there are many festivals with unique features in relation to the destinations where these events are held. These festivals are identified with those destinations and are a significant tool in the promotion of a region. In recent years, festivals have become an important attraction for destinations and with the increase in travel demand for the destinations organizing events and festivals, travel agencies start to give greater priority to festivals in their tour programmes.

Tourist guides are tourism employees who have one of the most important roles in the tourism industry. The main reason for this is that they are the people who are in constant and direct communication and interaction with tourists. They are the ones who make sure that tours and tourist trips have gone well. They are the ones who are effective in turning sudden negativities or setbacks into a positive direction and in completing tourists' experiences in a destination in the best way. They are also the ones who have a fundamental role in conveying the meaning of events and festivals to tourists. It is important for the quality of festival tours that tourist guides attend festivals, make observations and learn about the history of these kinds of events. In this context, the aim of the study reported in this chapter was to examine festivals organized in Turkey within the framework of cultural sustainability from the perspective of tourist guides.

Seven tourist guides from seven geographical regions of Turkey constituted the sample of this research. As a result of the interviews, the tourist guides claim that festivals have both positive and negative impacts on the natural and cultural environment of a region. Festivals are a source of cultural heritage and are intertwined with the concept of cultural sustainability. Considering cultural sustainability, the tourist guides assert some other notions such as customs and traditions, local food and local architecture. They also state that festivals have an important role and influence in ensuring cultural sustainability. When the literature is reviewed, it can be observed that the scope and number of studies on cultural sustainability are insufficient. In sum, according to the study findings, as a tool of culture, festivals are very significant for cultural sustainability, but it is possible to say that at present, the full potential of their contribution to achieving cultural sustainability is not being met as those involved in festival organization do not see the necessary importance of cultural sustainability.

References

Akın, M.H. (2007) Kültür Sosyolojisi. In: Alver, K. and Doğan, N. (eds) *Bir Kültür Bilimi Olarak Sosyoloji ve Kültür Sosyolojisi*. Hece Publishing, Ankara, pp. 85–96.

Assman, J. (2001) *Kültürel Bellek*. Translated by A. Tekin. Ayrıntı Publishing, İstanbul, Turkey.

Burke, P. (2006) *Kültür Tarihi*. İstanbul Bilgi Üniversitesi Publishing, İstanbul, Turkey.

Büyüköztürk, Ş., Çakmak, E.K., Akgün, Ö.E., Karadeniz, Ş. and Demirel, F. (2016) *Bilimsel Araştırma Yöntemleri*. Pegem Akademi, Ankara.

Cebeci, Ö.F. and Çakılcıoğlu, M. (2002) *Kültürel Sürdürülebilirlik, 10. Ulusal Bölge Bilimi/Bölge Planlama Kongresi*. Ekim, İstanbul, Turkey.

Coghlan, A., Buckley, R. and Weaver, D. (2012) A framework for analysing awe in tourism experiences. *Annals of Tourism Research* 39(3), 1710–1714. DOI: 10.1016/j.annals.2012.03.007.

Cudny, W. (2013) Festival tourism – the concept, key functions and dysfunctions in the context of tourism geography studies. *Geographical Journal* 65(2), 105–118.

Cudny, W., Korec, P. and Rouba, R. (2012) Resident's perception of festivals – a case study of Lodz. *Sociologia (Lisbon, Portugal)* 44(6), 704–728.

Ekiz, D. (2015) *Bilimsel Araştırma Yöntemleri*. Anı Publishing, Ankara.

Eroğlu, C. (2006) Türk kültürü ve temelleri. *Journal of Human Sciences* 8(1), 1–4.

Getz, D. (2008) Event tourism: Definition, evolution, and research. *Tourism Management* 29, 403–428. DOI: 10.1016/j.tourman.2007.07.017.

Getz, D. (2010) The nature and scope of festival studies. *International Journal of Event Management Research* 5, 1–47.

Getz, D. (2012) *Event Studies: Theory, Research and Policy for Planned Events*. Routledge, Abingdon, UK.

Getz, D. (2013) Festival tourism – the concept, key functions and dysfunctions in the context of tourism geography studies. *Geographical Journal* 65(2), 105–118.

Getz, D. and Page, S.J. (2016) Progress and prospects for event tourism research. *Tourism Management* 52, 593–631. DOI: 10.1016/j.tourman.2015.03.007.

Güvenç, B. (1985) *Kültür Konusu ve Sorunlarımız*. Remzi Publishing House, İstanbul, Turkey.

Kartarı, A. (2006) *Farklılıklarla Yaşamak: Kültürlerarası İletişim*. Ürün, Ankara.

Konuk, S. (2009) Bir planlama yaklaşim biçimi olarak kültürel sürdürülebilirlik kumkapi örneği. Master's thesis, İstanbul Technical University, İstanbul, Turkey.

Kucharska, W. and Kowalczyk, R. (2018) How to achieve sustainability?—Employee's point of view on company's culture and CSR practice. *Corporate Social Responsibility and Environmental Management* 26(2), 453–467. DOI: 10.1002/csr.1696.

Laing, J. (2018) Festival and event tourism research: current and future perspectives. *Tourism Management Perspectives* 25, 165–168. DOI: 10.1016/j.tmp.2017.11.024.

Larson, M. (2009) Joint event production in the jungle, the park, and the garden: metaphors of event networks. *Tourism Management* 30(3), 393–399. DOI: 10.1016/j.tourman.2008.08.003.
Mariani, M.M. and Baggio, R. (2012) Special issue: Managing tourism in a changing world: issues and cases. *Anatolia: An International Journal of Tourism and Hospitality Research* 23(1), 1–3. DOI: 10.1080/13032917.2011.653636.
Mignolo, W.D. (2011) *The Darker Side of Western Modernity: Global Futures, Decolonial Options*. Duke University Press, Durham, North Carolina.
Pelit, E., Demir, M. and Türkoğlu, T. (2018) Kültürel mirasin sürdürülebilirliği: turizm eğitimi veren kurum müfredatlari üzerine bir inceleme. *Turizm Akademik Dergisi* 5(2), 73–83.
Snepenger, D., King, J., Marshall, E. and Uysal, M. (2006) Modeling Iso-Ahola's motivation theory in the tourism context. *Journal of Travel Research* 45(2), 140–149.
UNWTO (World Tourism Organization) (2015) UNWTO Tourism Highlights, 2015 Edition. Available at: https://www.e-unwto.org/doi/pdf/10.18111/9789284416899 (accessed 20 January 2021).
Yıldırım, A. and Şimşek, H. (2013) *Sosyal Bilimlerde Nitel Araştırma Yöntemleri*. Seçkin Publishing, Ankara.
Yüksel, A. and Yüksel, F. (2004) *Turizmde Bilimsel Araştırma Yöntemleri*. Turhan Publishing House, Ankara.

Part 2: Green Events and Festivals

6 Iceland Airwaves Music Festival, a Green Festival? The Dissonance Between Stakeholders

Ian Jenkins[1]* and Robert Bristow[2]

[1]Háskóli Íslands/University of Iceland, Reykjavík, Iceland; [2]Westfield State University, Westfield, Massachusetts, USA

Introduction

The arts have long been a draw for tourism. In particular music festivals are a recognized contributor to tourism since they add to a region's identity and attract visitors. Iceland Airwaves music festival is one such event and has been gaining popularity for 20 years. Yet unlike other destinations that may have music festivals, as an island nation, Iceland must draw its visitors from thousands of kilometres away, and those who attend will typically fly and this contributes to the total carbon emissions for tourism significantly. While this may seem small, tourism does have an impact on global conditions. This chapter reports on the findings from the music festival attendees to ascertain the visitors' ranking of sustainability and offers strategies for music festival management to become more resilient.

> I would like to point out Iceland's uniqueness: Iceland and its pure untouched nature are synonymous. If that is lost our uniqueness is lost. Just as if Paris lost its fashion, New York lost its skyscrapers, Los Angeles its Hollywood. Our politicians must not get away with working against nature. We have to keep them on their toes. The damage could be too great, much greater than the short term profit.
>
> (Björk cited in Dibben, 2009, p. 1)

Festivals have become an extremely important component in any tourism strategy for a country or region (Liberato *et al.*, 2020). Iceland is no exception and perhaps it is even more important that festivals in Iceland are seen as successful given that its main economy is heavily reliant on the tourism product. Tourism in Iceland is now nearly as important as the fishing industry and its accelerated growth since 2010 has made major contributions to the economy (Jóhannesson and Huijbens, 2010; Ministry of Industries and Innovation and SAF, 2015). Tourism peaked in 2018 with 2.3 million visitors arriving via air or boat (Icelandic Tourist Board, 2021). The global pandemic in 2020 resulted in a global decline in tourism and Iceland is no exception.

It is somewhat ironic that much of the tourism growth could be attributed to what can be seen as negative imagery. The economic crash of 2008, which nearly bankrupted Iceland, and the eruption of Eyjafjallajökull in 2010, seem to have actually helped the image profile and increased the number of tourist visits to Iceland. Likewise, the active

*Corresponding author: jenkins@hi.is

DOI: 10.1079/9781789248685.0006

Fagradalsfjall volcano activity today (June 2021) draws thousands to observe nature's wrath.

Furthermore, Iceland has developed an appearance of a green destination founded upon the landscape and wilderness of the island and its perceived pristine environments (Ferðamálastofa, 2016). This is to some extent true but there are paradoxes, for example the Icelandic population is one of the highest carbon users in Europe (Statistics Iceland, 2018). Furthermore, 98% of all visitors arrive by plane and given the current debate on climate change this is not really seen as an environmentally sensitive way to travel to a destination (Brueckner and Abreu, 2017). Consequently Iceland's image of a green and pristine environment may be threatened by the dependence on fossil fuels by the visiting tourists.

The policy of encouraging festivals to attract more tourists has been a pillar of Iceland's tourist strategy (Ferðamálastofa, 2016) and is common to many national tourist boards. The most successful and largest festival in the country is Iceland Airwaves, which is based on Icelandic pop music (Gunnarsdóttir, 2016). Icelandic music is culturally distinctive and reflective of its cultural tenets (Dibben, 2009; Prior, 2015; Kerr *et al.*, 2017; Inspired by Iceland, 2019). For example, Prior (2015) suggests that the growth in the arts and music add another dimension to Iceland's reputation as a nature destination. So what appear to be two distinctly different attractions, the environment and culture, now have an interest in tourism development. The exponential growth of tourism on the island is now raising concerns related to its effects on the cultural and social integrity of Icelandic culture, together with concerns of damage caused to the environment (Jóhannesson and Huijbens, 2010; Sæþórsdóttir, 2010; Helgadóttir *et al.*, 2019).

As identified above, in terms of tourism development, Iceland has seen a spectacular rise in tourism (Jenkins, 2017). This growth has caused significant social, cultural and environmental impacts to the fragile fabric of this island, suggesting that Iceland would benefit from a planned approach to tourism sustainability (Sæþórsdóttir, 2010).

The Icelanders have certainly made the most of the tourism boom and there is much tourist development around the island and in particular Reykjavík, the capital. As with most capital cities, Reykjavík is where the majority of tourists visit and stay, usually then making journeys from the capital to the major tourist attractions around the island. The fastest transportation is by plane on Air Iceland from Reykjavík airport (a hub) and these flights connect to the other smaller towns, such as Akureyri, Ísafjörður and Egilsstaðir. There is one major road that surrounds the whole island and this too connects the other significant towns.

The festival programme in Iceland is quite full and is designed to run all year, with many festivals in summer which is when the majority of tourists visit Iceland. The weather at this time is comparably good with extended hours of sunlight too. The winter also has festivals and it is at this time that the Airwaves music festival is held. It is one of the largest (in terms of international visitor numbers) and one of the top ten festivals in Iceland and is truly international in terms of the audience that attend together with many of the famous bands that play at the festival (Jenkins, 2016; W. Larnach-Jones, Airwaves Managing Director, 19 March 2021, personal communication). However, much of the music is also Icelandic and many of the fans who attend come from Iceland while the internationally famous bands tend to draw more international guests.

The rest of this chapter follows this structure. First, we provide a brief examination of the uniqueness of island tourism and the policies facing island nations. This is followed by an introduction to the music festival called Iceland Airwaves (available at: https://icelandairwaves.is/ accessed 3 August 2021) and music festival attendees. Next the methods of examination used for a survey of attendees of the festival are described, in order to ascertain the visitors' ranking of to what extent Iceland Airwaves is considered a sustainable product. This is followed by the results of the survey. We conclude the chapter with a discussion of chief findings and prepare a list of ideas the multiple stakeholders might consider to make the event more resilient for all parties.

Islands Tourism, Policies and Music Festivals

Tourism on islands presents different environmental and sustainability issues from those of a tourist destination on the mainland (Modica and Uysal, 2016). The main issue is related to imports of goods and services and if the island is located in climate zones which are hostile to agriculture, then there is even more dependence upon imports. This is the case with Iceland, as it is just south of the Arctic Circle at latitude 66° N, which means that for much of the year the climate is not one where agricultural crops can be produced or livestock can be raised outside. These are cost factors and also environmental factors. Historically the threat of starvation and poor food sources were a problem for the Icelanders but this has now been circumvented by sea and air transport, bringing in the food produce and materials that are needed for the population. To source food locally is a challenge. There are hothouses, based upon geothermal heating, that do provide some produce, but the main growing season for outside crops is limited to 4 months of the year.

Both food sourcing and transport have large carbon footprint implications so events start with a high carbon footprint. That said, island environmental policies are being enacted to try and counter this such as offers to offset carbon in its airline tickets in order to address some of the carbon emissions; 40% of tourism's contribution to carbon in the world is from air travel (UNWTO-UNEP-WMO, 2008).

The extreme environment of Iceland has also created a high carbon usage of its population, especially related to road transport, which includes many tourists. In 2018, 42% of all greenhouse gas emissions were from road transport (MFTEANR, 2019). Many vehicles are large and have high carbon outputs (see Fig. 6.1). The reason is the terrain, especially the limited number of hard-surface roads. These vehicles further contribute to the large carbon footprint of the island.

The Icelandic Tourist Board's own research has identified that Iceland's nature includes the landscape and that there is a need to ensure that this is protected and enhanced (Ferðamálastofa, 2016). That said, when it comes to issues of sustainability multiple challenges exist for the island. The tourist board's policies must also be viewed within the wide debate of environmental impacts on the Icelandic landscape and ecosystems in Iceland. The European Union (EU) Land Use, Land Use Change and Forestry (LULUCF) Regulation 2018 now recommend forestry management and other resilient land usage as a means to reduce carbon in member countries (EU, 2018). This will be enacted by Iceland with the Icelandic government incorporating it into their National Plan (MFTEANR, 2019, 2020). The landscape of Iceland has been vigorously debated, especially whether to introduce tree and forestry management, as its natural ecosystems do not culturally or historically lend themselves to forestry management. Furthermore, there are heritage debates of whether introduction of

Fig. 6.1. Large 4×4 vehicle typical of many vehicles in Iceland. Photo courtesy of Ian Jenkins.

forests would be culturally at odds with Iceland's natural landscape.

The majority of tourists come to Iceland for the nature (Sutherland and Stacey, 2017) and this is connected to the marketing image of Iceland being an environmentally friendly destination. There are policies of the government to tackle climate change emissions which are laid out in their 2020 Carbon Plan (MFTEANR, 2019, 2020), yet it is difficult to identify similar policies relating to tourism and festivals. Given their economic importance, one would have thought that carbon reduction would be specifically incorporated within these sectors. Yet there are commercial organizations related to tourism who do advertise policies to encourage carbon reduction such as Islandica, an operator of tours in Iceland (ISLANDiCA, 2006). Nonetheless, while sustainability may be a concern for festival organizers, just how to define sustainability remains a challenge. For example, many stakeholders do not understand the mutual understanding between sponsors and attendees, and the lack of that understanding makes it difficult to obtain a sustainable music festival (Richardson, 2018).

Iceland Airwaves Festival

The Iceland Airwaves festival started in a small way at a hangar at Reykjavík city airport in 1999 and has become one of the top festivals in Iceland. The festival's goals are: (i) to promote the music of Iceland in Iceland and overseas; (ii) to advance the arts, culture and music life in Iceland; and (iii) to increase the number of tourists travelling to Iceland outside the summer high tourist season (Guðmundsdóttir *et al.*, 2020). Compared to the other festivals it has a large audience (9000 and 280 bands in 2015) and its main markets are geographically linked and reflect Iceland's tourist markets of European and North American tourists. In more recent years, there has been an increase of visitors from around the globe (W. Larnach-Jones, 19 March 2021, personal communication) from at least 60 different countries.

The festival mirrors tourism in Iceland, having had accelerated growth in visitors. Iceland has a strong cultural tradition of music and song and this festival is where the more successful Icelandic bands can have a wider audience to listen to their compositions. It is usually held in November each year and is designed to encourage tourists during this shoulder period (Sigurðardóttir, 2012). It runs for nearly a week and has a fringe festival, where many of the bands play in shops around Reykjavík, which further helps the local economy (Sigurðardóttir and Young, 2011), as tourists buy goods from the retail outlets where the bands are playing. It has always been closely connected to and sponsored by Icelandair and the festival connects with the airline to help fill airline seats.

In terms of environmental issues, few if any clear sustainable green policies could be identified. This seemed to change in 2016 when marketing material and other parts of the festival did begin to address the environmental aspects of the festival. The provision of reusable festival cups was one product introduced so that attendees could then use these in different venues for purchased drinks.

The biggest aspect of the environmental impacts has to be air travel to Iceland (Sharp *et al.*, 2016; Keeling, 2020). This seems to be an Icelandic challenge as in any island destination and is certainly a sensitive issue as tourism can only really exist in Iceland at the cost of using air transport.

Given Icelandic Tourist Board's marketing of Iceland as a pristine destination, one would expect that Iceland Airwaves would also be branded as an environmentally friendly festival as music festivals can help improve the image of a destination (Duarte *et al.*, 2018). Research by Háskóli Ísland in 2015 and 2016 investigated branding and environmental images of the festival (Jenkins, 2016). The researchers interviewed a sample of festival visitors to ascertain the visitors' perceptions of the festival and the brand it portrays. The research revealed that the festival's main focus was upon the exciting image of the festival and the cultural tenets of the music which Airwaves provided. It is evidenced that many festival goers see this as important but they did not identify the environment as part of the festival image. That said, the colour branding of the festival did suggest some environmental factors as the colours of green and blue together with purple were noted as representing the festival image (Jenkins, 2016), and could be tenuously linked to the environment of Iceland.

Most festival visitors stay for the week; however, they also use the other tourism products available and many go on excursions to destinations around Iceland. This also has an impact on the festival carbon footprint, as many of the excursions use vehicles that have very heavy carbon-burning engines (Kavan and Anděrová, 2020). These vehicles are large as they have been adapted to the extreme conditions in Iceland, which require large vehicles in order to navigate many of the minor roads which have no hard surfaces and frequently have many types of road hazards.

Music Festival Attendees

Music festivals are extremely popular and can attract large audiences; it has been suggested too that festivals and events have been one of the fastest-growing parts of the tourism industry, providing large economic benefits for communities (Sigurðardóttir and Young, 2011). This is especially true if the festivals are held in destinations with small populations. They can provide multiple opportunities for local businesses to provide for the festival visitors which increase the multiplier effect and ensure that leakages from the economic spend are low. Festivals then can be useful products to encourage tourism at a destination and, as argued above, given the age profile of audiences for pop music festivals, one might expect that the environment is protected and environmental policies are important and should be enacted and enhanced, especially in locations with environmental sensitivity.

Visitors are critical to any festival and the audience dictates the type of festival. Therefore the audience attitude should reflect the festival product. Given the emphasis now on climate change and environmental issues, it would be expected that the attitude of attendees should drive the product development of the festival. It is suggested that the younger the festival attendees are, the more aware they are of the environment and the damage festivals can have on the environment. Pop and rock music festivals are aimed at young audiences and it would be hypothesized that given the current seriousness of the climate debate, the audience should be environmentally motivated and have lifestyles to reflect that. However, that said, reviewing the more high-profile international music festivals such as Glastonbury and perhaps Montreux, the environmental extenuation factors are not clear. The evidence provided after a festival has finished shows images of large areas of waste and lack of care for the environment by the attendees. For example, reports and pictures of Glastonbury (York, 2015) show significant areas of the festival grounds covered by litter and discarded tents; this does not really resonate with what an environmentally friendly festival looks like. Yet Glastonbury does have environmental policies (Webster and McKay, 2016) and also encourages the attendees to use environmentally friendly products and transport. Some festivals like Paléo in Switzerland do have very clear environmental policies and seem to be environmentally friendly too. Websites and other sources provide insight to the level of environmental policy that a festival might provide (Paléo, 2021).

There are general policies and standards which music festivals should address relating to the environment and how to manage waste, such as ISO 20121 (created by the International Organization for Standardization (ISO, 2012)). There is also a discourse which suggests that festivals can influence the attendees to be more environmentally friendly through individual actions (Mair and Laing, 2012). There have been multiple calls to businesses to engage in sustainable practices (Jenkins and Schröder, 2013). The Global Sustainable Tourism Council (GSTC available at: https://www.gstcouncil.org/ accessed 5 August 2021) is one such body offering guidelines for industry leaders in tourism. Also many governments are employing tactics to increase the resilience in tourism. This top-down strategy is important. Here we explore the bottom-up approach to encouraging those organizations reluctant to take environmental concerns seriously to become more environmentally friendly, by providing evidence that the consumer desires sustainable practices.

Methods

To test this idea of a 'bottom-up' strategy to provide sustainable practices, a survey instrument was designed to measure the overall acceptance of sustainability issues at a music festival. The instrument was organized into four sections. The first, themed 'personal visit characteristics', inquired about personal practices that are considered sustainable and include recycling, purchasing organic food and overall concern about climate change. The second section asked about 'festival opinions' and sought information about the individual's perception of elements of sustainable practices at music festivals, in particular those concerning accommodation, awareness of environmental impacts and willingness to pay more for environmentally friendly events. In the third section, the survey solicited information about the festival product including merchandise, carbon footprint, and locally sourced food and beverages. Statements were listed in each section and these were assessed by the music festival participant on a five-point Likert scale, with a score of 1 for strongly agree to 5 for strongly disagree. An answer of 'not applicable' or 'not recorded' was coded missing for the analysis. The final section of the survey instrument collected standard demographic data such as gender, home country, age and occupation.

The instrument was dispatched at various venues during the 2016 Iceland Airwaves festival by a team of trained surveyors to obtain a random sample of participants at the music event. Then the data were entered into a database for statistical analysis. After final review of the data, a sample size of 206 was obtained and is described in the next section of the chapter. Very few refusals were recorded, but these are still reported.

Results

The purpose of the research is to identify to what extent is the Iceland Airwaves festival considered to be a sustainable product. The fundamental theme here is building resilience and promoting sustainability at Iceland's premier music festival. Given this underlying construct it is hypothesized that our music festival tourists may group similar concerns about sustainability to define the manner in which sustainable practices are part of their experience.

Basic demographic data were collected. Our sample was fairly evenly split in terms of gender. Of the respondents, 22% were from Iceland, about 50% were from Europe, 20% from North America (USA, Canada) and about 7% were from the UK. The remaining respondents were from outside these regions with a few guests from Asia. For ages, the crowd was young with 64% being under 35 years of age. A broad variety of occupations were recorded with those in education, either as a student or teacher, being the modal response (15% of the sample), and those in the arts or music careers representing nearly 10%. Occupations of the rest of the attendees surveyed varied tremendously, illustrating the broad appeal of Airwaves.

For personal visit characteristics, the statement 'I think that we need to do something about climate change' was the most agreed upon statement (with a score of 1.29/5.00). It also had a small standard deviation (SD) indicating strong agreement among the participants (SD = 0.494). The most disagreed statement was 'I think plastic bottles are not an environmental problem' illustrative of the general consensus against plastic bottles (4.38/5.00). Note, several of the survey statements were written in a negative manner (i.e. 'plastic bottles are *not* an environmental problem') to ensure the respondent did not automatically code every statement the same. Table 6.1 provides the means and standard deviations for these personal statements.

Festival opinions were solicited next and are summarized in Table 6.2. The table shows that 'I think festivals should try and be environmentally friendly' was rated high (1.58/5.00) and there was general agreement between the participants (SD = 0.619). There was slight disagreement for the statements 'I don't think Iceland is an environmentally friendly tourist destination' (3.79/5.00) and 'Festivals have very little environmental impact on the environment' (3.66/5.00). A wide response was attained to the statement 'I try to look for accommodation that is environmentally friendly' and the response was also somewhat neutral (i.e. neither agreed nor disagreed with the statement) (2.87/5.00; SD = 1.064).

In the next section of the survey, 'Products to be used at Airwaves', participants scored statements about products that Airwaves should be using. The statements relate to management strategies adopted by business stakeholders who cater for the festival attendees. Table 6.3 provides

Table 6.1. Personal visit characteristics.[a]

Statements	n	Mean score	SD
I think that we need to do something about climate change	206	1.29	0.494
I always recycle my rubbish at home	205	1.89	0.959
I try to use public transport rather than private cars	200	2.57	1.402
I don't think flying to destinations is contributing to global warming	198	3.77	1.201
I will pay more for environmentally sensitive products	203	2.21	0.906
I try to buy organic food	204	2.58	1.227
I think I should reduce the number of flights I use per year	193	3.44	1.131
I think plastic bottles are not an environmental problem	200	4.38	0.995

[a]Statements were scored using a five-point Likert scale with a score of 1 for strongly agree to 5 for strongly disagree. n, sample size (i.e. number of survey participants who responded to the statement); SD, standard deviation.

Table 6.2. Festival opinions.[a]

Statements	n	Mean score	SD
I try to look for accommodation that is environmentally friendly	185	2.87	1.064
I don't think Iceland is an environmentally friendly tourist destination	203	3.79	1.112
Festivals have very little environmental impact on the environment	203	3.66	0.980
I think festivals should try and be environmentally friendly	205	1.58	0.619
By attending Airwaves festival I am contributing to climate change	192	2.71	1.075
I wouldn't expect Airwaves to try and use the most environmentally friendly products	194	3.32	0.966
I would pay more for my ticket if the festival was more environmentally friendly	197	2.62	1.108

[a]Statements were scored using a five-point Likert scale with a score of 1 for strongly agree to 5 for strongly disagree. n, sample size (i.e. number of survey participants who responded to the statement); SD, standard deviation.

the descriptive statistics here. In general, festival participants desired merchandise from sustainable sources (1.87/5.00) and an 'Environmental policy for the festival' (1.94/5.00). Further, festival participants believed that buses should be used for transport between the airport and the festival as there was strong disagreement with the statement 'Airwaves shouldn't support use of buses from the airport to festival' (3.92/5.00).

In sum, these descriptive scores provide us with a ranking of sustainable practices as viewed by the music festival attendees in terms of personal practices, the perception of the festival and the use of sustainable products. Since there were 30 statements described above, we conducted a principal component analysis (PCA). PCA has been used to find similarities in tourism survey data as a data reduction procedure, since the statistics yield a smaller set of groups (Kozak, 2002; Bristow, 2020). For our research PCA was employed to help explain the common concerns of sustainability by those who visited the music festival. Cronbach's alpha for all the sustainability measures is 0.796, illustrative of a reliable and consistent survey instrument made up of Likert-style questions (Vaske, 2008).

The PCA with varimax rotation with Kaiser normalization was performed on the measures of sustainable practices for each of the three sections of the survey instrument. Multiple tests of significance were employed including the correlation matrix determinant, Bartlett's test of sphericity and the Kaiser-Meyer-Olkin (KMO) measure of sampling adequacy.

The eight personal visit characteristics were reduced to three factors exceeding an eigenvalue of 1.0 and after the varimax rotation, the three factors accounted for 57.06% of

Table 6.3. Products to be used at Airwaves.[a]

Statements	n	Mean score	SD
Environmentally sourced materials for those that cannot be recycled	203	2.16	0.962
Merchandising products should be from sustainable resources	204	1.87	0.683
Sell only locally sourced food at venues	205	2.38	1.147
Sell only locally produced beverages at venues	205	2.39	1.117
Measure the carbon footprint of the festival	202	2.19	0.961
Environmental policy for the festival	203	1.94	0.803
Carbon offset all visitors' footprints	190	2.67	0.937
Airwaves shouldn't support use of buses from the airport to festival	198	3.92	0.963
Should offer a reduction in ticket prices for all those that offset their air travel footprint	196	2.39	0.952
Ensure that all bands attend by the most environmentally friendly transport	200	2.45	0.944
Introduction of dry urinals in certain zones	196	2.33	1.006
Regular control of noise levels during concerts	190	2.99	1.217
Distribution of earplugs	198	2.22	1.028
Airwaves should plant trees to mitigate carbon emissions	199	2.38	0.997
Airwaves should support a global environmental organization	201	2.02	0.916

[a]Statements were scored using a five-point Likert scale with a score of 1 for strongly agree to 5 for strongly disagree. n, sample size (i.e. number of survey participants who responded to the statement); SD, standard deviation.

the variance (see Table 6.4). Correlations range from 0.000 to 0.518 suggesting a large amount of independence among the sustainability measures. The correlation matrix determinant of 0.364 is greater than 0.0001 indicating a low collinearity. Bartlett's test of sphericity was highly significant at $p = 0.000$ (Field, 2013). The KMO measure of sampling adequacy was 0.692 which is above the acceptable limit of 0.50 (Kaiser and Rice, 1974). This result yields a modest explanation, with slightly more than 40% of the variance being not explained. But given the number of personal visit characteristics (eight) compared with the sample size (n = 206) no reduction in the number of sustainable measures is recommended.

The first factor that we term home behaviour includes recycling at home, public transport use, and willingness to pay for environmentally sensitive products and organic food. These account for 24.82% of the variance and have similar loadings (0.539 to 0.767). These practices may be thought of as common sustainable practices at home. Factor 2 yielded loadings of 0.632 to 0.816 and reflected awareness of airline flight impact and plastic bottle waste and we consider these as strategies to reduce environmental impact. The second factor adds another 16.6% of the variance explained. The third factor yielded loadings of -0.654 to 0.766 and refers to awareness of the impact of climate change and the role of air flights.

The next PCA was prepared for the festival opinions and is found in Table 6.5. The seven festival opinions of personal visit measures were reduced to three factors exceeding an eigenvalue of 1.0 and after the varimax rotation, the three factors accounted for 62.7% of the variance. Loadings for the factors exceeded 0.62. The correlation matrix determinant of 0.410 indicates a low collinearity. Bartlett's test of sphericity was highly significant at $p = 0.000$. The KMO measure of sampling adequacy was 0.564, above the acceptable level of 0.50. Correlations range from 0.006 to 0.392 suggesting a large amount of independence among the sustainability measures.

Table 6.4. Factor analysis for personal visit characteristics.[a]

Personal visit characteristics	Eigenvalue	Variance explained (%)	Cumulative variance explained (%)
Factor 1 (home behaviour)	1.986	24.820	24.820
I always recycle my rubbish at home			
I try to use public transport rather than private cars			
I will pay more for environmentally sensitive products			
I try to buy organic food			
Factor 2 (reduce impacts)	1.324	16.545	41.366
I think I should reduce the number of flights I use per year			
I think plastic bottles are not an environmental problem			
Factor 3 (climate change)	1.255	15.689	57.055
I think that we need to do something about climate change			
I don't think flying to destinations is contributing to global warming			

[a]Kaiser-Meyer-Olkin (KMO) measure of sampling adequacy = 0.692; Bartlett's Chi square = 174.4, p = 0.000; Correlation matrix determinant = 0.364.

Table 6.5. Factor analysis for festival opinions.[a]

Festival opinions	Eigenvalue	Variance explained (%)	Cumulative variance explained (%)
Factor 1 (impacts)	1.929	27.55	27.55
I try to look for accommodation that is environmentally friendly			
I think festivals should try and be environmentally friendly			
By attending Airwaves festival I am contributing to climate change			
I would pay more for my ticket if the festival was more environmentally friendly			
Factor 2 (Airwaves' impact)	1.338	19.12	46.67
Festivals have very little environmental impact on the environment			
I wouldn't expect Airwaves to try and use the most environmentally friendly products			
Factor 3 (Iceland)	1.123	16.05	62.72
I don't think Iceland is an environmentally friendly tourist destination			

[a]KMO = 0.564; Bartlett's Chi square = 135.309, p = 0.000; Correlation matrix determinant = 0.410.

The first factor acknowledges the role of impacts of festivals and the willingness of attendees to pay more for more environmentally friendly festivals and accounts for 27.55% of the variance. The second group reflects the general acceptance that festivals have impacts and that Airwaves should use environmentally friendly products. This factor adds another 19.12% to the explained variance. The third factor shares the overall opinion of Iceland's role in being environmentally friendly and adds 16.05% variance so that together the three factors result in the total of 62.7% of variance in the model.

The last PCA is found in Table 6.6 and highlights the groupings uncovered for products that Airwaves could be using. Five factors are discovered and provide some useful insight about music festival participants' concern about sustainable purchasing. A total variance explained reached slightly more than 63%. Loadings for this group in the PCA begin at 0.472 and peak at 0.92. The correlation matrix determinant of 0.014 indicates a lack of collinearity. Bartlett's test of sphericity was highly significant at p = 0.000. The KMO measure of sampling adequacy was 0.755, well above the acceptable level of

Table 6.6. Factor analysis for products to be used at Airwaves.[a]

Products to be used at Airwaves	Eigenvalue	Variance explained (%)	Cumulative variance explained (%)
Factor 1 (carbon footprint)	3.109	20.73	20.73
Measure the carbon footprint of the festival			
Environmental policy for the festival			
Carbon offset all visitors' footprints			
Introduction of dry urinals in certain zones			
Airwaves should plant trees to mitigate carbon emissions			
Airwaves should support a global environmental organization			
Factor 2 (locally sourced food and beverages)	1.883	12.552	33.281
Sell only locally sourced food at venues			
Sell only locally produced beverages at venues			
Factor 3 (other items environmentally sourced)	1.659	11.058	44.34
Environmentally sourced materials for those that cannot be recycled			
Merchandising products should be from sustainable resources			
Factor 4 (transport and noise)	1.478	9.855	54.195
Airwaves shouldn't support use of buses from the airport to festival			
Regular control of noise levels during concerts			
Factor 5 (other)	1.322	8.813	63.008
Should offer a reduction in ticket prices for all those who offset their air travel footprint			
Ensure that all bands attend by the most environmentally friendly transport			
Distribution of earplugs			

[a]KMO = 0.755; Bartlett's Chi square = 632.02, p = 0.000; Correlation matrix determinant = 0.014.

0.50. Correlations range from 0.06 to 0.505 representing a large amount of independence among the sustainability measures.

The five factors obtained when festival attendees were asked 'Which of the following products should Airwaves be using?' reflect potential sustainable behaviour and several common themes. The first factor and accounting for 20.7% of the variance include elements of carbon footprints, management policies and strategies to offset carbon. Factor 2 combines the need for locally sourced food and beverages, and adds 12.5% of the variance explained. Non-food items are highlighted in factor 3 and account for an additional 11% of the variance.

The fourth factor explores the role of transport and noise (9.86% of variance). Factor 5 contributed the final 8.8% of the explained variance and encourages reduction of ticket prices for those who offset their carbon footprint, ensure that the bands attending the event also travel in the most environmentally sensitive manner and the provision of earplugs and yield a total variance explained at 63%.

In sum, the challenge of reducing 30 measures into similar factors that help us define what contributes to a sustainable festival has been accomplished using PCA. It seems to provide a clear indication of the need for change of the Airwaves festival to address environmental issues. The highest loaded factors in the three groups recognize the importance of a festival attendees' home behaviour (e.g. recycling), acknowledgement that they are contributing to the festival impact and are willing to pay to mitigate or offset that impact, especially for those practices that are tied to a carbon footprint.

Discussion

In this chapter we have collected information from a sample of attendees at a large music festival in Iceland. In particular we inquired about the personal practices of the tourists and their opinions of festival management and Iceland. These findings reflect a 'bottom-up' approach towards sustainable management for events and reflect the interests of multiple stakeholders.

Music festival attendees tend to be regular practitioners of sustainable habits at home, which supports the conjecture that younger age groups may be more environmentally aware. Other findings from Tables 6.1, 6.2 and 6.3 recognize the festival has multiple impacts, including attendees' own visitation to Iceland. Searching for types of accommodation that are green was not that important, but attendees were willing to reduce flights to decrease their carbon footprint. Since air transport is a major contributor to greenhouse gas emissions, this suggests a positive position for tourism. The lack of interest in green accommodation is modest, perhaps reflecting the nature of the festival itself: that is a place to crash.

However, it may well be that in Iceland it is difficult to identify accommodation that is environmentally friendly even though there is a grading system called Vakinn that visitors can use which includes environmental assessment in its award (available at: https://www.vakinn.is/en/ accessed 3 February 2022). Also how many tourists put this as a priority when booking a holiday? It is perhaps more the responsibility of the host to ensure that environmental factors are being met at a destination. The data shows that there appears to be an assumption and responsibility that Airwaves festival should have an environmental policy, should use environmentally friendly products and, in particular, plastic is seen as an important environmental issue to be addressed by the organizers.

The first factor in the PCA (factor 1) which yields the greatest explanation, and provides some interesting results, is that of personal visit characteristics (Table 6.4) composed of home behaviour, which included recycling, using public transport, willingness to pay more for environmentally sensitive products and purchasing organic food. These may be characteristics of the younger festival attendees and seem to support the hypothesis: *that younger individuals may well be more environmentally friendly*. Less important, while still significant, are the reduction of airline trips, plastic bottle use (both factor 2), and recognizing climate change (factor 3).

In Table 6.5, the first factor on festival opinions acknowledges the festival impacts including a general statement that festivals, and accommodation, should be environmentally friendly. The first factor also includes awareness that Airwaves contributes to climate change and the willingness of the tourist to pay more for the festival to be environmentally friendly.

The final PCA (Table 6.6 about products used at Airwaves) included several survey statements in the first factor, essentially all devoted to the carbon footprint of the festival and ways to mitigate it. The remaining factors for 'products' combine locally sourced food and beverages, and the reduction of practices that were perceived as detrimental to the environment. This may prove to be more difficult for Airwaves as imports to Iceland are a necessary part of island living which limits their options. However, checking supply chains and their environmental efficacy of the product would be appropriate for businesses. Further, food services could market themselves as environmentally green just as organic food has done in the past.

These results and findings need to be put into the context of island tourism and the difficulty of carbon footprinting, which will always be much higher given the geographical inaccessibility. The results should also be viewed within the context of the Icelandic government and Ferdamalastofa's policies relating to the environment and the need to reduce the Icelandic carbon footprint. In many cases the carbon measurement usually doesn't incorporate aircraft emissions (although domestic flights should be included); after COP21 (the 21st Conference of the Parties (COP) to the United Nations Framework Convention on Climate Change) it was noticeable that emissions were not included in the policies to reduce carbon. Therefore it would be unfair to reproach Airwaves and Iceland for ignoring the carbon emissions from tourism and festival visits. However, that is not to excuse them from making policies that would reduce carbon emissions. As identified in the earlier part of the chapter, currently the Icelandic carbon footprint is well above many other EU countries, including the now Brexit UK, which means that there is a need in Iceland to reduce carbon emissions even without air transport.

Perhaps key to this is the notion of offsetting by increasing or including a fee within the air ticket and Airwaves festival ticket to do this. This has been muted by Airwaves in 2021 (W. Larnach-Jones, 19 March 2021, personal communication) to be included in the 'Airwaves Ticket Plus'.

This chapter has demonstrated that Airwaves' management needs to review its approach to how it delivers the festival. A number of recommendations can be made:

1. There needs to be a policy to address the carbon footprint of air travel to the festival. Increased cost of tickets would help, related to carbon mitigation in line with EU, 2018 LULUCF regulations.
2. The Airwaves website should provide guidance relating to its festival environmental policies and actions, together with information for festival guests about accommodation and environmental travel.
3. Gaining ISO 20121 for the festival would validate an environmental approach.
4. Transport when attending the festival needs to be reviewed and use of public transport rather than cars needs to be encouraged.
5. The supply chain of products delivered at the festival needs to be ethical and appropriate where possible.
6. Products should be locally sourced but given the nature of island tourism this needs to be tempered with supply chain access.
7. Plastic product use needs to be diminished and discouraged.
8. A policy of waste management should be clearly visible for attendees and used by the festival.

Conclusion

In sum, tourists who attend the Iceland Airwaves festival tend to be individuals who exercise environmentally sound practices at home, are aware of their contribution to carbon emissions and are willing to pay more for the event to be more resilient. The data also shows that many sound environmental policies should really already have been adopted by the festival. Few if any attendees were negative about improving the environmental sustainability for the festival.

Given that most festival attendees flew to Iceland, at least their acknowledgement of an airline carbon footprint (Table 6.1) is an important step in the right direction. However, for an event to be sustainable, both the organizers and the participants need to recognize the overlapping responsibility of the task.

Events concomitantly have both positive and negative impacts on the society, economy and environmental realms of sustainability (ISO, 2012). Thus, Airwaves should follow the lead of the 2012 Olympics in London and assess their impact using ISO standards. ISO 20121 (*Event sustainability management systems – Requirements with guidance for use*; ISO, 2012) is designed to help planners address sustainability for everything from a music festival to the Olympics. These guidelines may be a challenge to achieve quickly, yet tourists need an awareness of the impact before any changes will occur (Bristow and Jenkins, 2017). Furthermore there are exemplars internationally of events and festivals with sound green festival policies and management, although it must be said these are not island festivals, namely Paléo in Switzerland and also one might consider Glastonbury in the UK too. Certainly sustainable festival management is now becoming a mandatory aspect of festivals rather than a green-washing approach. Also it should be noted that festival organizers have a moral responsibility to influence the festival attendees too and can further improve individual action of festival attendees through having green policies and products.

Festival organizers should embrace these strategies because our research has shown the tourists not only expect it for Iceland (Table 6.3), they also acknowledge they are contributors too (Table 6.2). Once on the island, additional research should explore the impact on travel and carbon emissions fuelled by the behaviour of participants and how changes to conditions and behaviour may well affect carbon emissions. Festival organizers now have the responsibility and the reason to behave sustainably; a resounding response of the 'bottom-up' approach to policy.

References

Bristow, R.S. (2020) Communitas in fright tourism. *Tourism Geographies* 22(2), 319–337. DOI: 10.1080/14616688.2019.1708445.

Bristow, R.S. and Jenkins, I. (2017) Travel behaviour substitution for a white-water canoe race influenced by climate induced stream flow. *Leisure/Loisir* 42(1), 25–46. DOI: 10.1080/14927713.2017.1403861.

Brueckner, J.K. and Abreu, C. (2017) Airline fuel usage and carbon emissions: determining factors. *Journal of Air Transport Management* 62, 10–17. DOI: 10.1016/j.jairtraman.2017.01.004.

Dibben, N. (2009) Nature and nation: national identity and environmentalism in Icelandic popular music video and music documentary. *Ethnomusicology Forum* 18(1), 131–151. DOI: 10.1080/17411910902816542.

Duarte, P., Folgado-Fernández, J.A. and Hernández-Mogollón, J.M. (2018) Measurement of the impact of music festivals on destination image: the case of a Womad festival. *Event Management* 22(4), 517–526. DOI: 10.3727/152599518X15264726192451.

EU (European Union) (2018) *Land Use and Forestry Regulation for 2021–2030*. Available at: https://ec.europa.eu/clima/policies/forests/lulucf_en (accessed 3 March 2022).

Ferðamálastofa (2016) Ferðamálastofa – Icelandic Tourist Board International Visitors in Iceland Summer 2016. Available at: https://www.ferdamalastofa.is/static/files/ferdamalastofa/Frettamyndir/2017/januar/sunarkonnun/sumar-2016-islensk.pdf (accessed 5 November 2021).

Field, A. (2013) *Discovering Statistics Using IBM SPSS Statistics*. Sage, Thousand Oaks, California.

Gunnarsdóttir, N. (2016) Top 10 Festivals in Iceland. *Guide to Iceland*. Available at: https://guidetoiceland.is/best-of-iceland/the-top-10-festivals-in-iceland (accessed 3 March 2022).

Guðmundsdóttir, E.R., Sigurðardóttir, M.S. and Torfason, M.Þ. (2020) Iceland Airwaves sem samskiptavettvangur og farvegur hugmynda. *Íslenska Þjóðfélagið* 11(1), 41–59.

Helgadóttir, G., Einarsdóttir, A.V., Burns, G.L., Gunnarsdóttir, G.Þ. and Matthíasdóttir, J.M.E. (2019) Social sustainability of tourism in Iceland: a qualitative inquiry. *Scandinavian Journal of Hospitality and Tourism* 19(4–5), 404–421. DOI: 10.1080/15022250.2019.1696699.

Icelandic Tourist Board (2021) Numbers of Foreign Visitors. *Icelandic Tourist Board, Research & Statistics*. Available at: https://www.ferdamalastofa.is/en/recearch-and-statistics/numbers-of-foreign-visitors (accessed 3 March 2022).

Inspired by Iceland (2019) Things to do, FESTIVALS. Available at: https://visiticeland.com/festivals-in-iceland (accessed 17 March 2021).
ISLANDiCA (2006) 5 Ways How to Reduce Your Carbon Footprint While in Iceland. Available at: https://www.viajesislandia.com/en/reduce-carbon-footprint (accessed 16 March 2021).
ISO (International Organization for Standardization) (2012) *Systèmes de Management Responsable Appliqués à l'activité Événementielle – Exigences et Recommandations de Mise En Oeuvre [Event Sustainability Management Systems – Requirements with Guidance for Use]*. ISO, Geneva, Switzerland.
Jenkins, I.S. (2016) *Airwaves Branding, Report*. Department of Geography and Tourism, Háskóli Íslands, Reykjavík.
Jenkins, I.S. (2017) Case study Iceland: climate change and tourism sustainability and its effects on Icelandic coastal destinations. In: Jones, A.L. and Phillips, M. (eds) *Global Climate Change and Coastal Tourism: Recognizing Problems, Managing Solutions and Future Expectations*. CAB International, Wallingford, UK, pp. 147–158.
Jenkins, I. and Schröder, R. (eds) (2013) *Sustainability in Tourism: a Multidisciplinary Approach*. Springer Science + Business Media, Media, Heidelberg, Germany.
Jóhannesson, G.T. and Huijbens, E.H. (2010) Tourism in times of crisis: exploring the discourse of tourism. *Current Issues in Tourism* 13(5), 419–434.
Kaiser, H.F. and Rice, J. (1974) Little jiffy, mark IV. *Educational and Psychological Measurement* 34(1), 111–117.
Kavan, J. and Anděrová, V. (2020) Impacts of increased tourism on polar environment – case studies from Svalbard and Iceland. *Czech Polar Reports* 10(1), 59–68.
Keeling, D.J. (2020) Transport geography in Iceland. *Journal of Transport Geography* 89, 102875. DOI: 10.1016/j.jtrangeo.2020.102875.
Kerr, B.A., Birdnow, M., Hallaert, J., Alexander, K., Malmsten, R. *et al.* (2017) Creativity and innovation in Iceland: individual, environmental, and cultural variables. *Gifted and Talented International* 32(1), 27–43. DOI: 10.1080/15332276.2017.1397903.
Kozak, M. (2002) Comparative analysis of tourist motivations by nationality and destinations. *Tourism Management* 23(3), 221–232. DOI: 10.1016/S0261-5177(01)00090-5.
Liberato, D., Costa, E., Liberato, P. and Ribeiro, J. (2020) The role of events and music festivals in urban tourism: case study. In: Rocha, Á., Abreu, A., de Carvalho, J., Liberato, D., González, E. *et al.* (eds) *Advances in Tourism, Technology and Smart Systems: Proceedings of ICOTTs 2019. Smart Innovation, Systems and Technologies*, Vol. 171. Springer, Berlin, pp. 537–549. DOI: 10.1007/978-981-15-2024-2_47.
Mair, J. and Laing, J. (2012) The greening of music festivals: motivations, barriers and outcomes. *Journal of Sustainable Tourism* 20(5), 683–700.
MFTEANR (2019) *Iceland's National Plan*. Government of Iceland, Ministry for the Environment and Natural Resources, 25 October. Available at: https://www.stjornarradid.is/library/02-Rit--skyrslur-og-skrar/Iceland%20National%20Plan%202020.pdf (accessed 16 March 2021).
MFTEANR (2020) *Iceland's 2020 Climate Action Plan*. Government of Iceland, Ministry for the Environment and Natural Resources, October. Available at: https://www.government.is/library/01-Ministries/Ministry-for-The-Environment/201004%20Umhverfisraduneytid%20Adgerdaaaetlun%20EN%20V2.pdf (accessed 3 February 2022).
Ministry of Industries and Innovation and SAF (Icelandic Travel Industry Association) (2015) Road Map for Tourism in Iceland. Ministry of Industries and Innovation and SAF, 3 October. Available at: https://www.stjornarradid.is/media/atvinnuvegaraduneyti-media/media/Acrobat/Road-Map-for-Tourism-in-Iceland.pdf (accessed 4 February 2022).
Modica, P. and Uysal, M. (eds) (2016) *Sustainable Island Tourism: Competitiveness and Quality of Life*. CAB International, Wallingford, UK. DOI: 10.1079/9781780645421.0000.
Paléo (2021) Our environmental policy. Available at: https://yeah.paleo.ch/en/environmental-policy (accessed 16 March 2021).
Prior, N. (2015) 'It's a social thing, not a nature thing': popular music practices in Reykjavík, Iceland. *Cultural Sociology* 9(1), 81–98.
Richardson, N. (2018) Corporate social responsibility or sustainability in music festivals. *International Journal of Organizational Analysis* 27(5), 1257–1273. DOI: 10.1108/IJOA-03-2018-1368.
Sæþórsdóttir, A.D. (2010) Planning nature tourism in Iceland based on tourist attitudes. *Tourism Geographies* 12(1), 25–52. DOI: 10.1080/14616680903493639.

Sharp, H., Grundius, J. and Heinonen, J. (2016) Carbon footprint of inbound tourism to Iceland: a consumption-based life-cycle assessment including direct and indirect emissions. *Sustainability* 8(11), 1147.

Sigurðardóttir, K. (2012) *TónlistarháTíðin Iceland Airwaves*. Markaðssetning og skipulag, BS ritgerð, Líf-og umhverfisvísindadeild, Háskóli Íslands, Iceland, 33 bls.

Sigurðardóttir, M.S. and Young, T. (2011) *Towards Creative Iceland: Building Local, Going Global*. Icelandic Ministry of Finance, Reykjavík.

Statistics Iceland (2018) Iceland has the highest per capita CO2 emissions from the economy. Available at: https://www.statice.is/publications/news-archive/environment/carbon-dioxide-emission-per-capita/ (accessed 4 February 2022).

Sutherland, D. and Stacey, J. (2017) Sustaining nature-based tourism in Iceland. *OECD Economics Department Working Papers* No. 1422. Organisation for Economic Co-Operation and Development (OECD), Paris.

UNWTO-UNEP-WMO (2008) *Climate Change and Tourism: Responding to Global Challenges*. United Nations World Tourism Organization (UNWTO), United Nations Environment Programme (UNEP) and World Meteorological Organization (WMO), Madrid.

Vaske, J.J. (2008) *Survey Research and Analysis: Applications in Parks, Recreation and Human Dimensions*. Venture Publications, Glossop, UK.

Webster, E. and McKay, G. (2016) From Glyndebourne to Glastonbury: the impact of British music festivals. *The Impact of Festivals Project*. University of East Anglia. Available at: https://ueaeprints.uea.ac.uk/id/eprint/59132/1/Festival_Report_online.pdf (accessed 16 March 2021).

York, M. (2015) Glastonbury's rubbish: going against the green ethos ruins it for everyone. *The Guardian*, 1 July. Available at: https://www.theguardian.com/music/2015/jul/01/glastonburys-rubbish-green-ethos-ruin-festival-worthy-farm-tents (accessed 14 March 2021).

7 Music, Green Marketing and Sustainability Festivals: the Case of the Rock in Rio Tourist Event

Annaelise Fritz Machado[1]*, Bruno Sousa[1], Joice Lavandoski[2], Laurentina Vareiro[1] and Victor Figueira[3]

[1]Polytechnic Institute of Cávado and Ave (IPCA), Barcelos, Portugal; [2]Universidade Federal do Estado do Rio de Janeiro (UNIRIO), Rio de Janeiro, Brazil; [3]Polytechnic Institute of Beja (IPB), Beja, Portugal

Introduction

Tourism is a multifaceted and geographically complex activity that increasingly generates new (and different) market segments with different individual interests (Sousa *et al.*, 2021). Festival tourists undoubtedly seek appealing, exclusive and memorable music sensory impressions in the festival tourism settings. Events and music festivals have evolved over the last few decades because of the competitiveness between music-producing destinations. Therefore, musical events and festivals are a strong tourist attraction in the international context. The recent pandemic has naturally changed the strategic positioning of some of these musical events. However, in the 'pre-pandemic' context, the issue of sustainability was heavily debated by scholars of marketing and tourism. In an exploratory perspective, this chapter seeks to understand the sustainable initiatives present at the event 'Rock in Rio' as a world reference in musical events. Events have contributed to controlling the seasonality of tourist destinations and as they can attract large numbers of tourists, bringing income that boosts the local economy, event tourism has become an expanding tourism sector. Stakeholders have been following the movements of the market and realized that organizations have grown when they include a capable sustainability plan in their strategic planning, in order to meet consumer expectations as well as the global guidelines in the United Nations (UN) Sustainable Development Goals (SDGs) (UN Brasil, 2015) and green marketing trends. This chapter takes a qualitative perspective (i.e. using semi-structured interviews) and content analysis to compile a list of best practices in managing music festivals with environmental concerns in a sustainable way.

Green marketing and events trends

According to Shabani *et al.* (2013) the first definition of green marketing was presented in 1979 by Henion, who assumed that green marketing involved marketing programmes and activities that focused on the environment-friendly sectors or the same green market

*Corresponding author: annaelisefritz@yahoo.com.br

 Festival and Event Tourism: Building Resilience and Promoting Sustainability (eds A. Sharma *et al.*)
DOI: 10.1079/9781789248685.0007

approach. Coddington (1990) argues that green marketing is marketing actions that identify environmental stewardship as a business growth opportunity and business development responsibility. Environmental concerns and consumer demand for environmentally friendly products or services have led to the emergence of a new marketing philosophy called green marketing. The great forces behind green marketing are the demand for green products coupled with concern for the environment. On the other hand, companies seek to balance their sales and profit objectives with social and environmental concerns (Peneda de Oliveira and Sousa, 2020). However, according to these previous authors, the main focuses back then were the size of the green market and the 'profile' of the green consumer. Some scholars even say that the consumer profile was the only area of interest in studying the greening of the consumer. Green marketing is a broader concept that covers many more aspects such as consumer goods, industrial goods and services as well (Chen, 2009; Peneda de Oliveira and Sousa, 2020).

Chen (2009) developed a theoretical framework which shows that the image of the brand, satisfaction and trust act as boosters that increase brand equity 'Green' (Mourad *et al.*, 2012). It is understood that it is a way of articulating the relationships between consumers, companies and the environment (Lange *et al.*, 2018). The Green Seal, a non-profit environmental standard development and certification organization, was created by governments and non-governmental organizations (NGOs) from several countries that, driven by the concern of consumers with the degradation of the environment, elaborated a set of rules that regulate the trade of products. The Green Seal administers the International Standards Organization (ISO) ISO 14001 Environmental Certification, which is awarded to a company certifying that it follows the rules and implements the indicated processes and, therefore, it is an organization that has environmental responsibility. This operates worldwide and indicates that a company that acts in accordance with these standards can be considered socially responsible, when presenting the three main international certifications: (i) ISO 14001, OHSAS 18000, SA 8000, ISO 9001 (Quality); (ii) ISO 14001 (Environment); and (iii) OHSAS 18001 (Safety and Occupational Health) (Kanedo *et al.*, 2015). To further complement these there is the business ISO 20121, the sustainable event management certification standard (ISO, 2012).

Events impact the environment and the economy at various levels, including: (i) water consumption; (ii) solid waste production; (iii) energy consumption; (iv) threats to biodiversity; (v) employment rate; and (vi) selection of suppliers. Society is now environmentally aware and gives greater importance to the care of the environment than it did in the past. Therefore the objective of this study is to develop a model of the behaviour of the ecological consumer in order to know the motivations that influence consumption of organic products or sustainable events and services. In this context, the ISO 20121 sustainability management system helps the organization to establish and achieve realistic goals, as well as execute policies and procedures for applying good practices (SGS, 2020). An example of this, in Brazil, is the Certification Exhibition Management (CEM), whose certification is the responsibility of the International Association of Exhibitions and Events (IAEE), represented by NewEvents Global. Portugal observes the existence of the 'Most Sustainable Event' certification process, awarded by SGS, which has structural indicators to ISO 20121. However, it is observed that this certification process is not always simple, as there is resistance from the consumers themselves, in addition to technical difficulties of application by companies and the expense of inserting new methodologies that are required.

Rock in Rio

Rock in Rio is a recurring music festival originating in Rio de Janeiro, Brazil. It later branched out into other locations such as Lisbon, Madrid and Las Vegas. Eight incarnations of the festival have been held in Rio de Janeiro (in 1985, 1991, 2001, 2011, 2013, 2015, 2017 and 2019), eight in Lisbon (in 2004, 2006, 2008, 2010, 2012, 2014, 2016 and 2018), three in Madrid (in 2008, 2010 and 2012) and one in Las Vegas (in 2015). Brazilian entrepreneur and advertiser Roberto Medina was responsible for the inception and organization of the festival, as well as moving the 2004 edition to Lisbon, while controversially keeping the brand 'Rock in Rio'. In 2011, Rock in Rio returned to its original location, Rio de Janeiro, with a new line-up of singers and groups. Rock in Rio is one of the largest music festivals in the world, with 1.5 million people attending the first event, 700,000 attending the second and fourth, about 1.2 million attending the third, and about 350,000 people attending each of the eight Lisbon events (Medina, 2019).

With the end of the military dictatorship in Brazil, the musical knowledge that was restricted to artists at the time began to be explored by more people. In this context, in 1985, Roberto Medina had a visionary and daring idea, to create an event that could innovate through a communication platform with unique experiences, which would bring visibility to Brazil, economic gains and possible investors and entrepreneurs to Rio de Janeiro. It created the brand Rock in Rio, putting the country on the map of major international shows because of the strength of the rock music (Anubis, 2014). The name chosen for the event was also designed to be accessible to everyone and highly visible. So, Medina thought of the word rock coming from the English 'rock and roll', and ending with the word Rio coming from Rio de Janeiro. That's how the biggest music event in the world was founded (Ribeiro, 2015).

Rock in Rio's Marketing Director, Agatha Arêas, recalls that the festival began to be designed in 1984, as a tool to move not only the economy, but also the cultural scene and self-esteem in Rio de Janeiro and Brazil. She said:

> The country was leaving the dictatorship very faint. With the censorship of artists, Rio de Janeiro, which has always been a natural source of creativity, was a faded city at that time. The Rock in Rio project was and is really to rescue the city's self-esteem, promote joy and allow the whole country to dream again.
>
> (Agatha Arêas cited in Época, 2017; and cited in Medina, 2019)

As Brazil at the time was not prepared to organize large festivals, several entrepreneurs did not believe in the potential of the event conceived by Medina, who was trying to seek support in the USA. He went to New York alongside producer Luiz Oscar Niemeyer and businessman Oscar Ornstein, where they spent 70 days holding various meetings with internationally relevant singers, including The Rolling Stones, The Who, Deep Purple, The Smiths, Depeche Mode, Michael Jackson and even bands that were no longer active, like Led Zeppelin and Pink Floyd (Anubis, 2014). But nothing was achieved there. Faced with this reality, Medina left for Los Angeles and looked for businessman Lee Solters, who accepted the event and disclosed it to 70 journalists who released the news that Brazil would organize the greatest music festival of all time (Anubis, 2014).

The event Rock in Rio has as its central value to make a difference in the world, so it supports various social and environmental causes. Sustainability, promotion of social actions, environmental protection and reforestation are part of the concept 'For a Better World' at Rock in Rio. Even before the slogan was officially adopted by the festival in 2001, the social proposal was already intrinsic to the event. Since 1985, the number of initiatives has been growing in different areas and places (Época, 2017). The nomenclature launched in 2001 'For a Better World' is to improve the world with the participation of all, by making daily changes, and to meet the promise of sustainability set out in the ISO 20121 agreement the festival organizers developed a sustainability plan for the music festival. This plan includes social, environmental and economic measures that improve the work performance of the organization, sponsors and suppliers. Environmental measures include: (i) calculating the carbon footprint; (ii) a comprehensive waste management plan to reduce, reuse and recycle as much waste as possible, and has achieved an overall average rate of 70% for recycling and waste recovery, which includes the donation of materials at the end of the event; and (iii) the award for sponsors and suppliers with the best sustainable practices in the City of Rock (Época, 2017). It is interesting to note that much more than just linking its brand to the cause, Rock in Rio allocates part of its income to social projects (Teixeira *et al.*, 2018), an item that will be discussed later in the chapter.

Case Study: Green Trends and Rock in Rio

A new awareness of how human activity is damaging the biosphere has given organizations an impetus to develop strategic plans that favour enforcement actions intended to guide the productive use of renewable resources without diminishing the productivity and environmental quality of the world. In this context, the search for sustainability in an organization's business can be reproduced through what is known as the triple bottom line or sustainability tripod, bringing a new look to companies, government and society as a whole (Elkington, 2001). The triple bottom line or sustainability tripod is an accounting framework to evaluate performance and achieve a balance between three dimensions: economic, environmental and social (Trigo and Senna, 2016; Machado *et al.*, 2021). In the corporate world, companies have been challenged to assume postures that avoid environmental degradation and ensure their survival and financial sustainability, which makes them competitive. Green marketing, when companies develop and advertise products or services that are presumed to be environmentally safe, appears to be a support mechanism and is consistent with environmentally sound practices, meeting the needs of those customers who want a more sustainable lifestyle that causes less damage to nature (Souza, 2013).

Green marketing not only adds value to the final product, but also brings benefits to consumers in terms of social and motivational aspects. In the face of these inferences, the UN, considering different national circumstances, capacities and levels of development, and respecting national policies and priorities, created the SDGs that represent global priorities for the 2030 Agenda signed by more than 190 countries, including Brazil and Portugal. In 2019, Rock in Rio adopted the SDGs as a guide for its sustainability plan. Some of the sustainable projects carried out by Rock in Rio that work with the SDGs are as follows.

(a) Amazônia Live

The campaign Amazônia Live Rock in Rio is proof of the great concern for sustainable attitudes. Organizers have launched an alert on the climate issue and have started a major movement in defence of the planet. Rock in Rio set out to plant 1 million trees in the deforested areas of the Xingu springs and, with everyone's help, the goal has been exceeded: 4 million new trees have been planted in the Amazon (Beegreen, 2017). To plant the trees, a mix of local seeds called muvuca were used, a technique chosen and improved by the Instituto Socioambiental (ISA) to reproduce the natural process of the forest. The technique uses no-till planters, which open a slit in the soil, and seeds are planted directly into the ground in the final position, as this is the best method for most types of trees. Trees have been planted in the Xingu-Araguaia ecoregion of moist forests (for the 2017 season). The main purpose of this project is to make everyone aware of the importance of preserving and restoring the environment, especially the Amazon rainforest. According to Doare, a partner of the Rock in Rio that takes care of the donations to the Amazônia Live, you can make a donation of R$5.90 to plant a tree or even R$529.00 for planting 100 (Teixeira *et al.*, 2018). A space was placed on the event's official website where people could donate money online to plant trees. All money raised through the website was sent to the Brazilian Fund for Diversity (FUNBIO), responsible for the financial management of the planting (Beegreen, 2017).

(b) Solar trees

Through the Amazônia Live project, five 'solar trees' were installed in Rock City at the site of the music festival that, besides being islands of rest, allowed participants to recharge their mobile phones (Beegreen, 2017) using a technology that transforms solar energy into electrical energy (Teixeira *et al.*, 2018). The invention is capable of capturing solar energy through photovoltaic films installed on the sheets. Each tree contains ten outlets that provide free energy, raising public awareness of the importance of conscientious consumption and environmental responsibility (Beegreen, 2017).

(c) Waste management

Waste management was implemented at Rock in Rio for the first time in 2008 in Lisbon. The waste management plan is based on three main sustainability issues: (i) economic (with the reuse of materials); (ii) environmental (by reducing the amount of waste sent to landfill, preventing the generation of waste and producing organic compost for reforestation); and (iii) social (with the donation of materials and surplus food) (Beegreen, 2017).

(d) Rock in Rio – responsible attitude

Not only does the organization responsible for the event want it to be sustainable, it also requires sustainable attitudes from its sponsors. Since 2008, the organization rewards the partners for their sustainable actions and incorporating effectiveness with the sustainability plan. The event's sponsors are: (i) Itaú; (ii) City Hall of Rio de Janeiro; (iii) Postos Ipiranga; (iv) Doritos; (v) Heineken; and (vi) Coca-Cola (Beegreen, 2017). The organization has been investing in an extensive plan to reduce the event's carbon dioxide (CO_2) emissions at an international level. To this end, it provided a manual for companies working at the event on the application of sustainable practices during the assembly and disassembly of equipment and structures. It was the first festival to recycle 100% of the waste produced, both in Brazil and in Portugal (Trigo and Senna, 2016).

The attitude of Rock in Rio event organizers is to encourage the use of public transport that meets the goals of British Standard (BS) 8901, which was created specifically for the Olympic Games in London, in 2012. In line with the guidelines of BS 8901, from an environmental point of view, organizers of Rock in Rio demonstrated a commitment to be a '100R' event, that is to recycle the waste produced at all stages of the event from the assembly, going through its realization until demobilization (post-event) (Trigo and Senna, 2016). In the social and economic aspects of the guidelines of BS 8901, it is observed that Rock in Rio Lisboa 2008 and the 2010 event motivated the community to become more involved, thus generating local socio-economic development. The standard also contributed to a lower consumption of water and energy during the event in 2012. Other actions were important:

> 3,200 young people with basic education in Rio de Janeiro;
>
> 100 classrooms in pacified deprived areas of Rio de Janeiro;
>
> 10 music rooms in Rio de Janeiro;
>
> 28 projects financed through UNESCO;
>
> 1 school in Tanzania;
>
> Projects in 43 countries – Plan International Child Reach;
>
> 1 health center in Maranhão;
>
> 14 sensory rooms in Portugal that support thousands of young people with disabilities;
>
> 760 solar panels in schools – 15 years generating income for social projects;
>
> 445,500 young people saved resources valued at 1.8 million euros;
>
> 19,925 meals and 37,000 sandwiches were donated in Portugal and Las Vegas;
>
> 2,200 instruments donated to 150 NGOs [non-governmental organizations] in Brazil;
>
> 350,260 usd [US$350,260] invested in 80 scholarships in music education in Portugal;
>
> + 97 million reais invested;
>
> More than 200 entities supported and more than 56,400 people benefited/year.
>
> (Rockinrio.com, 2019, p. 33)

It is observed that the organizers of Rock in Rio are engaged in developing actions that respect the environment, develop and welcome local people with regard to social responsibility, and follow global standards of sustainability, which provides financial health to sustain the business.

Final Considerations

Leisure is a phenomenon that moves millions of people around the world and is one of the main drivers of a global economy. Every year, due to frequent changes in the leisure and tourism environment, leisure fosters competition between and within destinations (e.g. music festivals), making it increasingly a multifaceted and geographically complex activity, where different services are requested and provided at different stages, from origin to destination (Ferreira and Sousa, 2020). The event Rock in Rio, known nationally and globally, is a multicultural music event which has adopted a management plan based on the pillars of sustainability, including environmental, economic and social measures. The initiatives in the sustainability plan support environmentally correct practices, while being economically viable, and socially fair and culturally accepted, so that the event has achieved worldwide certification in these areas. Rock in Rio has assumed its mission in the 21st century, to help transform the planet, not only through music and entertainment, but also through environmental, social and cultural causes.

The analysis of the data presented here makes it possible to say that:

1. There has been a change in the behaviour of the Rock in Rio event organizers (in terms of business logic) with regard to the company's adequacy to the environmental, economic and social sustainability certification policies, including labour norms and standards.
2. The inclusion of green marketing strategies demanded of companies in the corporate world to avoid environmental degradation and ensure their survival and financial sustainability, has driven innovation at Rock in Rio making it possible to develop a strategic plan for the company.
3. Rock in Rio event organizers have integrated the triple bottom line (sustainability tripod) into their business model, balancing environmental, economic and social issues with actions that make it possible to respect the environment and future generations.
4. The events' sustainability plan allows work on projects that are in alignment with the 17 SDGs suggested by the UN, such as Amazônia Live, installing solar trees, waste management control and the requirement that sponsors of Rock in Rio are required to demonstrate sustainable attitudes – as presented here in the case study.
5. Sponsors and stakeholders involved in the Rock in Rio event are required to develop creative and sustainable actions in their contributions during the event, which contribute to the event receiving ISO 20121 certification, a world award for sustainable events.
6. There has been an increase in economic movement to the city of Rio de Janeiro with an increase in the flow of people in hotels, restaurants and tourist spots, enabling the tourist trade to become more successful.
7. There has been an improvement in the basic and tourist infrastructure of the city of Rio de Janeiro, the economic impact of which has created around a billion direct and indirect jobs and the movement of 700,000 people, of which 65% are tourists (Tanabe, 2019).

In general, it can be said that the objectives defined for this study were satisfactorily achieved, showing that consumers perceive and react positively to Rock in Rio's involvement in sustainable practices that contribute to achieving the SDGs. It is concluded that the Rock in Rio event stimulates the dynamics of the destination and promotes sustainable actions linked to the environment and social responsibility, the basis of green marketing. A limitation of this research is that the literature search yielded few articles published in journals of great relevance to the subject area of the Rock in Rio event and there are limited financial resources to monitor events that may still happen in the future. For future studies on Rock in Rio, we see the relevance of an investigation with quantitative data, since there is a scarcity of this sort of research.

References

Anubis, M. (2014) Rock In Rio, a história da primeira edição do maior festival de música do mundo. Available at: https://cwblive.com/rock-in-rio-a-historia-da-primeira-edicao-do-maior-festival-de-musica-do-mundo/ (accessed 12 August 2021).

Beegreen (2017) Confira as ações sustentáveis do Rock in Rio 2017. Available at: https://beegreen.eco.br/confira-as-acoes-sustentaveis-do-rock-in-rio-2017/ (accessed 3 January 2021).

Chen, Y.S. (2009) The drivers of green brand equity: green brand image, green satisfaction, and green trust. *Journal of Business Ethics* 93(2), 307–319. DOI: 10.100710551-009-0223-9.

Coddington, W. (1990) How to green up your marketing mix. *Advertising Age* 61(1), 30.

Elkington, J. (2001) *A Teoria dos Três Pilares*. Tradução de Patrícia Martins Ramalho. Markron Books, São Paulo, Brazil.

Época (2017) Rock in Rio apoia ações sociais e ambientais. Available at: https://epoca.globo.com/Especial-Publicitario/Rock-in-Rio/noticia/2017/09/rock-rio-apoia-acoes-sociais-e-ambientais.html (accessed 6 January 2021).

Ferreira, J. and Sousa, B. (2020) Experiential marketing as leverage for growth of creative tourism: a cocreative process. In: Rocha, Á., Abreu, A., de Carvalho, J., Liberato, D. and González, E. (eds) *Advances in Tourism, Technology and Smart Systems. Smart Innovation, Systems and Technologies*. Springer, Berlin, pp. 567–577. DOI: 10.1007/978-981-15-2024-2_49.

ISO (International Organization for Standardization) (2012) *Systèmes de management responsable appliqués à l'activité événementielle – Exigences et recommandations de mise en oeuvre* [*Event sustainability management systems – Requirements with guidance for use*]. ISO, Geneva, Switzerland.

Kanedo, A.T., Correia, D.N. and Gonçalves, M.M. (2015) *Marketing Verde Como Diferencial Das Empresas Sustentáveis*. Trabalho de Curso de Administração de Empresas. Fundação de Ensino 'Eurípides Soares da Rocha', mantenedora do Centro Universitário Eurípides de Marília – UNIVEM, Marília, Brazil.

Lange, K. de O., Trindade, L. de O. and Marques, J. da S. (2018) Sustentabilidade e marketing verde nas empresas. *RCA (Revista Científica da Ajes)* 7(14), 38–46.

Machado, A.F., Vareiro, L., Sousa, B., Figueira, V. and Lavandoski, J. (2021) Music festivals and green marketing and tourism: the case Rock in Rio. Paper presented at ICITED 21 – International Conference in Information Technology & Education, 15–17 July. Sao Paulo, Brazil.

Medina, R. (2019) Rock in Rio anuncia ganhadores do Prêmio Rock in Rio Atitude Sustentável 2019. Available at: https://cdn.rockinrio.com/wp-content/uploads/2020/03/rock-in-rio-2019_-rock-in-rio-anuncia-ganhadores-do-premio-rock-in-rio-atitude-sustentavel-2019_.pdf (accessed 3 February 2021).

Mourad, M., Serag, E. and Ahmed, Y. (2012) Perception of green brand in an emerging innovative market. *European Journal of Innovation Management* 15(4), 514–537. DOI: 10.1108/14601061211272402.

Peneda de Oliveira, C. and Sousa, B.M. (2020) Green consumer behavior and its implications on brand marketing strategy. In: Naidoo, V. and Verma, R. (eds) *Green Marketing as a Positive Driver Toward Business Sustainability*. IGI Global, Hershey, Pennsylvania, pp. 69–95. DOI: 10.4018/978-1-5225-9558-8.ch004.

Ribeiro, T.D. (2015) O modelo de negócio do Rock in Rio – uma perspetiva exploratória. Dissertação apresentada ao mestrado de Gestão e Estudos da Cultura, ISCTE – Instituto Universitário de Lisboa, Lisbon, Portugal.

Rockinrio.com (2019) Rock in Rio: Plano de Sustentabilidade Versão 1.1. Available at: http://cdn.rockinrio.com.br.s3.amazonaws.com/manual_uploads/Rock_in_Rio_2019_Plano_de_Sustentabilidade_versao_1_1.pdf (accessed 3 January 2021).

SGS (2020) ISO 20121 – Sustainable Events. Available at: https://www.sgsgroup.com.br/pt-br/sustainability/management-and-compliance/organization-and-events/iso-20121-sustainable-events (accessed 2 January 2020).

Shabani, N., Ashoori, M., Taghinejad, M., Beyrami, H. and Fekri, M.N. (2013) The study of green consumers' characteristics and available green sectors in the market. *International Research Journal of Applied and Basic Sciences* 4(7), 1880–1883.

Sousa, B.B., Magalhães, F.C. and Soares, D.B. (2021) The role of relational marketing in specific contexts of tourism: a luxury hotel management perspective. In: Rodrigues, P. and Borges, A.P. (eds) *Building Consumer-Brand Relationship in Luxury Brand Management*. IGI Global, Hershey, Pennsylvania, pp. 223–243. DOI: 10.4018/978-1-7998-4369-6.ch011.

Souza, V. (2013) Marketing Verde: consciência ou estratégia? 26 April. Available at: https://administradores.com.br/artigos/marketing-verde-consciencia-ou-estrategia (accessed 3 February 2021).

Tanabe, A. (2019) Rock in Rio movimenta economia, hotelaria e turismo no Rio de Janeiro. *Economia do turismo*, 9 October. Available at: https://diariodoturismo.com.br/rock-in-rio-movimenta-economia-hotelaria-e-turismo-no-rio-de-janeiro/ (accessed 4 December 2021).

Teixeira, B.C., Silva, E.R.D., Cardoso, I.M., Oliveira, P.A., Moraes, R.L.G. de *et al.* (2018) *Rock in rio: uma análise de marketing*. Etec, Atibaia, Brazil.

Trigo, A.G.M. and Senna, J.S.M. (2016) Sustentabilidade em eventos: características, motivações e análise de eventos sustentáveis. *Congresso Nacional de Excelência em Gestão*, Inovarse. Available at: https://www.inovarse.org/sites/default/files/T16_080.pdf (accessed 3 January 2021).

UN (United Nations) Brasil (2015) Transformando Nosso Mundo: a Agenda 2030 para o Desenvolvimento Sustentável. Available at: http://nacoesunidas.org/pos2015/agenda2030/ (accessed 5 April 2021).

8 The Effectiveness of Organizational Absorptive Capacity on Green Innovation Performance among Event SMEs: a Comparative Study of Malaysia and Singapore

Rupam Konar[1]*, Kashif Hussain[1], Vinayaraj Mothiravally[2] and Kandappan Balasubramanian[1]

[1]Taylor's University, Subang Jaya, Selangor, Malaysia; [2]Dimension International College, Singapore

---

Introduction

In a competitive business environment, service firms struggle unceasingly to improve their services, competences and performance through creating, integrating and enhancing new and existing capabilities. With rapid technological advancements, the service-product life cycle is getting shortened, and the necessity for service innovation and organizational intelligence skills is becoming more evident to service-based businesses. Alongside the adaptation of innovative services, with the emergence of social pressure for reduced carbon footprints for events and mega-events, event service providers are forced towards becoming sustainability oriented. Absorptive capacity is defined as a firm's ability to identify and acquire external knowledge (new information) and use it or apply it to commercial ends in order to gain performance. Due to gained popularity, various event companies have started operating and organizing events on a large scale. These firms are mainly small and medium-sized enterprises (SMEs). However, with due pressure from government bodies towards transforming operations with reduced carbon footprints, these SMEs are further trying to align their operations towards the national key areas and international Sustainable Development Goals (SDGs). Beside the pressure from the negative environmental impact caused by event operations, green innovation on products, services and process have become an important business strategy for a firm to achieve sustainable development (Agbejule *et al.*, 2004; Chang, 2011). This is driven by customer requirements and competitive pressures. Due to the massive pollution problems, the public's awareness of environmental protection has been significantly enhanced and people are willing to buy environmentally friendly products and services with a higher price. As the green market is continually growing, developing green products and services is necessary for a firm to obtain brand differentiation and competitive advantage (Pujari *et al.*, 2003; Chang and Chen, 2013). If an event management firm fails to adopt green practices, it

*Corresponding author: Rupam.Konar@taylors.edu.my

DOI: 10.1079/9781789248685.0008

may cause the firm to lose current and potential customers and it may become hard for the firm to survive in hypercompetitive scenarios (Butler, 2008).

Fukey and Issac (2014) highlighted that event staff do not have sound and complete knowledge of and expertise in environmental management. However, innovating environmentally friendly operations is not an easy task, given the capacity of the firm and their internal resources. Previous research on Malaysian event management firms' green practices found that employees are aware of environmental responsibility and are willing to perform green practices (Bouchon *et al.*, 2017). However, environmentally friendly practices would not be performed well due to the lack of a supporting system. SMEs located in Kuala Lumpur and Singapore generally lacked knowledge of environmental management (Kasim, 2009). Yusof and Jamaludin (2014) found that the main barriers for Malaysian enterprises to go green are lack of green experts, and lack of resources in terms of manpower and green equipment. Additionally, a number of studies (De Marchi, 2012; Cainelli *et al.*, 2015; Marzucchi and Montresor, 2017) indicated the relevance of environmental innovations through public-private collaborations in the event SMEs. However, it is less known how those firms could acquire knowledge based on their internal capabilities to absorb the external knowledge, and transform such relevant knowledge into offering services or products that meet the environmental concerns. Hence, this chapter focuses on carrying out a comparative study of the green innovation performance of event management firms in Malaysia and Singapore, as previous researchers have highlighted environmental management practices and green practices are closely linked to a firm's innovativeness and this is still considered new for the event management industry in these two countries.

Theoretical Background and Hypothesis Development

Green innovation performance

From any organizational perspective, innovation is defined as introduction of new services or products which bring changes to the organization to a certain degree (radical or incremental) (Ashok *et al.*, 2016; Hussain *et al.*, 2016). In the environmental era, there is constant pressure from internal bodies that firms should integrate such ideas to protect the environment. Thus, green innovation within services and products has become quintessential for a firm's business management. As Chang and Chen (2013) highlighted, efficient management can increase a firm's performance, competitive advantage, and create value for the long term. In order to avoid environmental damage and lead the way for future sustainability, green innovation is crucial for any firm, as it defines the new and improved green practices, processes, techniques and systems (Beise and Rennings, 2005). Such important changes within the operations of any SMEs in the event industry could significantly transform their current operations towards green operations. These practices could include, for example: (i) energy-saving initiatives while organizing events; (ii) pollution prevention measures reducing environmental damage; (iii) use of biodegradable materials; (iv) green production designs; (v) use of recyclable materials; and (vi) waste management facilitation. The event management firms could utilize such changes to attract environmentally conscious customers, as presently customers around the world want and expect to purchase more environmental services and products during events.

From the supplier's perspective, such demand from customers can be fulfilled by implementing green innovation practices in their service operations. Authors such as Hart (1995) and Porter and Reinhardt (2007) suggested that green innovation performance may increase a firm's performance through maximizing productivity of their used resources. Attracting new customers and increased service yields will further promote the corporate image and provide competitive advantage. SMEs in the event industry across Malaysia and Singapore are aligned with the government direction towards GoDigital by 2022 and have been advised to boost green-tech within the services they provide. This initiative also includes other countries in the Association of Southeast Asian Nations (ASEAN) region under the development plan. Such initiatives can certainly boost their green innovation services while reducing environmental impacts dramatically. Sustainable core services related to innovation and different strategies to implement them should be a firm's aim in greening their activities. However, the green initiatives adopted by these event management firms are very ordinary in nature rather than innovative, and many researchers have highlighted that lack of absorptive capacity could be the reason for such slackness (Duchek, 2013; Duan *et al.*, 2020). Meaningful collaboration and the ability to learn from each other could play a crucial role in elevating the pace of innovation in new green technologies.

Absorptive capacity

The concept of 'absorptive capacity' has been preliminarily defined by Cohen and Levinthal (1990) as the ability of a firm to identify the valuable knowledge, assimilate the knowledge and exploit it from its dynamic environment. Absorptive capacity has been further highlighted as a company's ability not only to gain knowledge, but also to make use of such knowledge effectively in order to innovate and gain competitive advantage in the dynamic market (Lichtenthaler, 2009). Chen *et al.* (2009) also highlighted that absorptive capacity is essential for organizations to enhance their innovation performance as absorptive capacity has a positive relationship with the innovation performance. In terms of the influence towards a firm's performance, it has been confirmed that absorptive capacity could, directly and indirectly, positively influence the organization's performance and assist it to create a competitive advantage (Lane *et al.*, 2001; Chen *et al.*, 2009). Therefore, if a company wants to prove successful and viable in the long run in terms of being a green operation, it is very important that they have the absorptive capacity to go green as such absorptive capacity shows positive results in relation not only to green innovation performance, but also to total firm performance (Chen *et al.*, 2015).

Chen *et al.* (2014) proposed that green absorptive capacity should be measured by five items: (i) the organizational structure's ability to understand, analyse and interpret information from the external environment; (ii) the organization's ability to recognize, value and select suitable external environmental knowledge which matches its operation; (iii) the ability for the environmental knowledge to be communicated across the organization; (iv) the organization's ability to combine existing environmental knowledge with new environmental knowledge; and (v) the organization's ability to commercialize new environmental knowledge.

Absorptive capacity and green innovation performance

Service innovation affects a firm's performance, or put another way how well a firm performs indicates how good the firm is at innovation in services (i.e. how good the firm is at developing new and improved services which are valuable for their customers). Service innovation can be studied from the macro aspect and the micro aspect. In the macro view, the focus is on the impact of service innovation on the firm's performance as a whole; while in the micro view, the focus is on the impact of service innovation on detailed measurement of the firm's performance (Lin, 2013; Chong and Zhou, 2014). More researchers have conducted studies of the macro view. Based on the impact of green innovation on a firm's performance, previous researchers have different opinions. One group of researchers believe that green innovation performance has a positive influence on a firm's performance (Chen *et al.*, 2015; Lee and Min, 2015; Huang and Li, 2017). Table 8.1 provides a summary of findings from the literature on the positive relationship between absorptive capacity, green (service) innovation and a firm's performance.

Zhang *et al.* (2019) measured a firm's performance using sales growth and net profit as variables. In terms of operational performance, Pujari (2006) found that enough and effective green integration practices positively assist an organization to achieve resource optimization and reliable production, which save time. Hence given the background knowledge described above we propose the following hypothesis, H1: absorptive capacity positively relates to green innovation performance.

Table 8.1. Summary of findings from the literature on the relationship between absorptive capacity, green (service) innovation and a firm's performance.

Findings	Reference	Type of industry	Overall summary of findings
Positive relationship between (green) absorptive capacity and green (service) innovation	Hashim *et al.* (2015)	Construction industry	All the authors' studies found that absorptive capacity is positively related to organizational service innovation. Chen *et al.* (2015) further stated that green absorptive capacity indirectly influences green innovation performance.
	Chen *et al.* (2009)	Manufacturing industry	
	Chen *et al.* (2015)	Electronics industry	
Positive relationship between (green) absorptive capacity and a firm's performance	Chen *et al.* (2009)	Manufacturing industry	All the authors' studies found that absorptive capacity improves a firm's performance. Lin *et al.* (2017) demonstrated that absorptive capacity improves a firm's green competitive advantage. Chen *et al.* (2015) stated there is an indirect and positive effect from green absorptive capacity on a firm's performance through either green dynamic capabilities or green innovation performance.
	Lane *et al.* (2001)	International joint ventures	
	Chen *et al.* (2015)	Electronics industry	
	Lin *et al.* (2017)	Education	
Positive impact of green (service) innovation in relation to a firm's performance taken as a whole	Chen *et al.* (2015)	Electronics industry	All the authors supported that green innovation performance positively influences firm performance.
	Huang and Li (2017)	Information and communication technology industry	
	Lee *et al.* (2016)	Firms in Korea	
	Chang (2018)	Service and manufacturing companies	

Research Methodology

Data collection and sampling

This research focuses on the event management industry, specifically SMEs located across Malaysia and Singapore. Highly dependent on tourism, hospitality and events, these two countries have created a significant impact within the ASEAN region over the past decade. This current study's targeted population was limited to the employees, owners and other related stakeholders working with SME firms located in Singapore and Malaysia. The data collection for this survey was undertaken based on non-parametric purposive sampling, where the samples were self-selected working at different levels in firms which started providing sustainable and green services during organizing events. With the rigid criteria set for data collection, combining two countries, we managed to collect 213 valid responses from 237 completed survey questionnaires. With the valid response rate of 89.5%, the data analysis was performed for both measurement and structural model using WarpPLS 7.0 software.

Measurements made from the questionnaire

In the questionnaire the research instrument items for the absorptive capacity were adapted from Albort-Morant *et al.* (2018) and Chaudhary and Batra (2018) while the green innovation firm performance items were adapted from Albort-Morant *et al.* (2016, 2018). The study used a five-point Likert scale with a score ranging from 1 (for strongly disagree) to 5 (for strongly agree) to measure the questionnaire items. The absorptive capacity (AC) dimension was represented by potential absorptive capacity (PAC) using nine items to assess it, and another construct, realized absorptive capacity (RAC), used ten items, assessing the firm's newly acquired knowledge. The green innovation performance (GIP) includes seven measurement items, as overall the model of this study is represented in first-order construct levels. Considering the measurement items were obtained mainly from two sources and self-reported methods are subject to common method bias (CMB) (Podsakoff *et al.*, 2003), a priori and post hoc strategies were performed by us to minimize this potential issue. This study used partial least squares (PLS) path modelling to test the hypothesis.

Data analysis

The data analysis for this study was divided into a measurement model and a structural model. Using PLS path modelling, variance-based structural equation modelling (SEM) was performed to get the results for data analysis. Assessing the measurement model enables examination of the validity and reliability of the measures of theoretical constructs – the outer model – and estimation of the relationship among the studied constructs – the inner model.

Results

The demographics of the respondents to the questionnaire showed that 62.2% were male and 37.8% were female with combined responses from Malaysia and Singapore. In regards to their age, the majority of the respondents were aged between 19 and 35 years (61.3%), followed by 27.5% aged between 36 and 49 years, and 11.2% aged 50 years and above. In respect of the department of the event management firm in which respondents worked, the majority of the respondents belonged to the operational department (53.2%), but those from the administrative department (18.8%) and marketing department (19.4%) were well represented.

Measurement model

In order to ensure the reliability and validity of the constructs in the model, measurement model evaluation was carried out first. The following evaluation followed the rule of thumb for a reflective measurement model as summarized by Konar *et al.* (2018): (i) internal consistency reliability; (ii) convergent validity; and (iii) discriminant validity. Internal consistency reliability requires Cronbach's alpha (α) > 0.7, Jöreskog's ρc > 0.7, and Dijkstra-Henseler's ρA > 0.7 (Henseler *et al.*, 2016). The measurement model shows α, ρA and ρc for all constructs were above 0.7, with the reliability value ranging from 0.857 to 0.959 (see Table 8.2). Thus, the results indicate that all the constructs used have satisfactory internal consistency reliability. Convergent validity has mainly been evaluated by measuring the outer loading of indicators and average variance extracted (AVE). The outer loading of indicators represents the variance of the items according to the latent variable, which requires a latent variable able to explain at least 50% variance of each indicator, and requiring that the indicator's outer loading should be above 0.708 (see Table 8.3).

Table 8.2. Internal consistency reliability.

Construct	Alpha (α)	Rho (ρA)	Rho (ρc)
Potential absorptive capacity	0.857	0.903	0.902
Realized absorptive capacity	0.870	0.919	0.897
Green innovation performance	0.949	0.951	0.959

Table 8.3. Indicator's validity and reliability.

Items	AVE[a,b]	CR[a]	Loadings[b]
Potential absorptive capacity (PAC)	0.518	0.887	
PAC1			0.912
PAC2			0.675
PAC3			0.879
PAC4			0.856
PAC5			0.735
PAC6			0.743
PAC7			0.764
PAC8			0.856
PAC9			0.879
Realized absorptive capacity (RAC)	0.589	0.859	
RAC1			0.856
RAC2			0.879
RAC3			0.900
RAC4			0.851
RAC5			0.824
RAC6			0.895
RAC7			0.907
RAC8			0.810
RAC9			0.856
RAC10			0.879
Green innovation performance (GIP)	0.612	0.899	
GIP1			0.856
GIP2			0.735
GIP3			0.743
GIP4			0.764
GIP5			0.856
GIP6			0.879
GIP7			0.900

[a]AVE, Average variance extracted; CR, composite reliability.
[b]Critical values: AVE = 0.50; indicator loadings = 0.708.

Table 8.4. Discriminant validity – the heterotrait-monotrait (HTMT) ratios for this study.[a,b]

	PAC	RAC	GIP
PAC			
RAC	0.869		
GIP	0.891	0.871	

[a]Use of the shaded boxes is the standard procedure for reporting HTMT ratios.
[b]PAC, potential absorptive capacity; RAC, realized absorptive capacity; GIP, green innovation performance.

In order to ensure the constructs in the model are empirically distinct from each other, discriminant validity must be evaluated. In this study, two measures were used to assess discriminant validity: (i) heterotrait-monotrait (HTMT) ratio; and (ii) cross-loading. An HTMT ratio < 0.85 means 95% confidence to consider the reflective constructs used in the study have a true correlation with conceptually different constructs (Henseler *et al.*, 2015). For cross-loading, an indicator's loading on the respective construct should be greater than the loading value shown on all other constructs in the row and column (Hair *et al.*, 2019). Table 8.4 shows the HTMT ratios for this study: the highest HTMT ratio is for construct 'green innovation performance' at 0.891, which is lower than the critical value 0.90. Therefore, the results indicated that the constructs used in this study have a true correlation.

Structural model

The coefficient is considered to be statistically significant at a certain level which is measured by the *t*-value and *p*-value. The critical values are: (i) a *t*-value equal to 1.65, a *p*-value equal to 0.1 indicate significance at the 90% level; (ii) a *t*-value equal to 1.96, a *p*-value equal to 0.05 indicate significance at the 95% level; and (iii) a *t*-value equal to 2.57, a *p*-value equal to 0.01 indicate significance at the 99% level. Path coefficients were considered as significant when the level was at 95% and above (Hair *et al.*, 2019).

In order to test the significance level of the mentioned hypothesis across two contexts, a multi-group analysis (MGA) was conducted to identify the difference between Malaysia and Singapore; it is further important to check the *p*-value and *t*-value among the two models. The results show: (i) the *t*-value = 2.695 in the Malaysian context, which is the lowest value when compared to *p*-value = 0.005; (ii) the *t*-value = 2.983 which belongs to 'absorptive capacity > green innovation performance' in the Singapore context, which was significant at the 99% level; and (iii) the *p*-value and *t*-value for the rest were significant at the 99% level. Therefore, this study has at least 95% confidence and above to state that all independent variables have a positive effect on its respective dependent variable, according to both the f^2 value and Q^2 values. In terms of f^2 effect size, the minimum effect size value is 0.141 and the largest effect size value is 1.327; based on the rules of thumb, the results could be summarized as: 'absorptive capacity' has a medium effect size on 'green innovation performance'. In terms of Q^2 value, the minimum effect size value is 0.614 (Fig. 8.1) and the largest one is 0.653 (Fig. 8.2), which means that the Q^2 values for all endogenous constructs are above 0.35. Therefore, the results indicate that the items in each exogenous construct have large predictive relevance on the respective endogenous construct.

This study has used four criteria to test model fit, which are: (i) the standardized root mean square residual (SRMR); (ii) the unweighted least squares discrepancy (d_{ULS}); (iii) the geodesic discrepancy (d_G); and (iv) the normal fit index (NFI) to test model fit (Konar *et al.*, 2018). The rule of thumb is SRMR < 0.08, d_{ULS} < 95% bootstrap quantile (H195 of d_{ULS}): (critical value < 0.05) (Henseler *et al.*, 2016), d_G < 95% bootstrap quantile (H195 of d_G): (critical value < 0.05) (Henseler *et al.*, 2016), and NFI value > 0.90 (Byrne *et al.*, 2008). Table 8.5 shows that the structural model of this study demonstrated an SRMR value of 0.061 and 0.060, which are both less than the critical value 0.08, which indicates that the structural model of this study has a perfect model fit. In terms of d_{ULS} and d_G, the results for the saturated model were 0.0001 (d_{ULS}) and 0.0038 (d_G), and for the estimated model 0.0002 (d_{ULS}) and 0.0064 (d_G), which are all less than the critical 0.05. Thus, it also confirmed that the structural model is considered as 'well fitting' based on the obtained values of d_{ULS} and d_G. The value of NFI was 0.93, which is above the criterial value 0.90. Therefore, it also

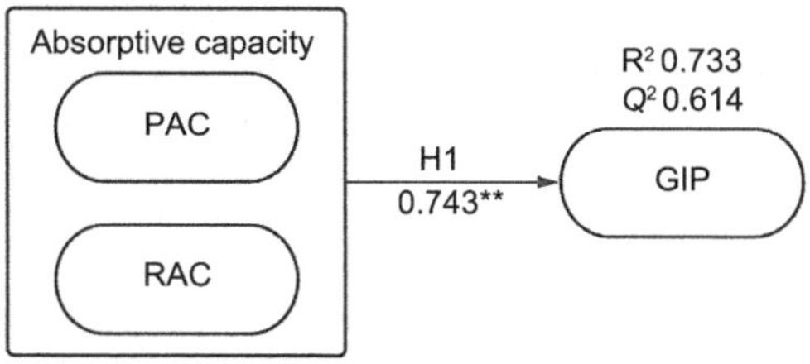

Fig. 8.1. Green innovation performance (GIP) in the Malaysian context. H1, hypothesis; PAC, potential absorptive capacity; RAC, realized absorptive capacity.
Critical values **2.57 *1.96.
R^2 Coefficient value; Q^2 Predictive relevance.

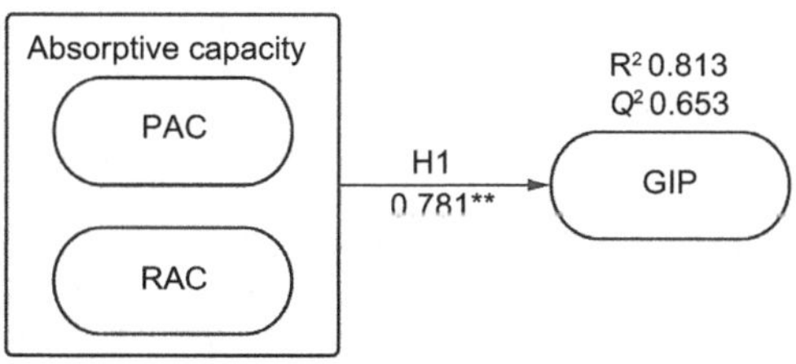

Fig. 8.2. Green innovation performance (GIP) in the Singaporean context.
Critical values **2.57 *1.96.
R^2 Coefficient value; Q^2 Predictive relevance.

Table 8.5. Exact fit tests.

Criteria[a]	Saturated model	Estimated model
SRMR	0.061	0.060
d_{ULS}	0.0001	0.0002
d_G	0.0038	0.0064
NFI	0.93	0.93

[a]SRMR, standardized root mean square residual: critical value (< 0.08); d_{ULS}, unweighted least squares discrepancy: critical value (< 0.05); d_G, geodesic discrepancy: critical value (< 0.05); NFI, normal fit index: critical value (> 0.90).

confirmed that the structural model of this study is well fitting. Based on the above analysis, it could be confirmed that the structural model of this study is well fitted. The results of first order constructs demonstrate that green absorptive capacities positively influence green innovation performance. Therefore, hypothesis H1 (that absorptive capacity positively relates to green innovation performance) for the overall model and contextual models are both supported.

Discussion and Conclusion

This research has referred to the measurement of absorptive capacity by Albort-Morant *et al.* (2016, 2018) and Chaudhary and Batra (2018), identifying the absorptive capacity of event management-based SMEs towards green innovation performance, as the ability: (i) to identify, value and obtain suitable external green knowledge which matches the daily operation of the firm; (ii) to combine the newly obtained green knowledge with existing green knowledge; (iii) to commercialize the green knowledge; and (iv) to communicate the green knowledge across each department in the event management firm. According to the framework proposed in existing studies, absorptive capacity has been used many times as a measure of green innovation performance. For event SMEs in the service industry across Singapore and Malaysia, therefore, service performance is always considered as a key measure for any service innovation. Based on the information above, this study proposed a conceptual framework that green absorptive capacity has a positive influence on green innovation performance. The results from this study demonstrated a confidence level of 99%, confidence to state that green absorptive capacity has a positive influence on green innovative performance. This finding is consistent with the findings of Chen *et al.* (2009) and Lane *et al.* (2001), which confirmed absorptive capabilities positively influence a firm's performance and competitive advantage. In relation to the 'green' component, this finding is further supported by Chen *et al.* (2015) and Ali *et al.* (2017), who confirmed absorptive capacity positively influences a firm's performance and helps firms to gain a green competitive advantage. This is aligned with Homburg and Pflesser (2000) and Gabler *et al.* (2015) who confirmed the positive influence of absorptive capacity on organizational performance, marketing performance and financial performance.

In conclusion, event management SMEs located in Malaysia and Singapore have the knowledge for conducting green innovation performance; however, the firms could not utilize the existing knowledge to go green well due to the lack of a related professional workforce, resources and technology to support it. More than that, the event management firms currently performed well with regard to their green absorptive capacity but they also need to continually identify, obtain, combine and communicate green knowledge in the firms in order to support dynamic green service development.

References

Agbejule, A., Fernández, M. and d'Espiney, S. (2004) Approaches to environmental value analysis of products, processes, and services. *Management of Environmental Quality: An International Journal* 15(2), 111–130. DOI: 10.1108/14777830410523062.

Albort-Morant, G., Leal-Millán, A. and Cepeda-Carrión, G. (2016) The antecedents of green innovation performance: a model of learning and capabilities. *Journal of Business Research* 69(11), 4912–4917. DOI: 10.1016/j.jbusres.2016.04.052.

Albort-Morant, G., Leal-Rodríguez, A.L. and De Marchi, V. (2018) Absorptive capacity and relationship learning mechanisms as complementary drivers of green innovation performance. *Journal of Knowledge Management* 22(2), 432–452. DOI: 10.1108/JKM-07-2017-0310.

Ali, F., Hussain, K., Konar, R. and Jeon, H.M. (2017) The effect of technical and functional quality on guests' perceived hotel service quality and satisfaction: a SEM-PLS analysis. *Journal of Quality Assurance in Hospitality & Tourism* 18(3), 354–378. DOI: 10.1080/1528008X.2016.1230037.

Ashok, M., Narula, R. and Martinez-Noya, A. (2016) How do collaboration and investments in knowledge management affect process innovation in services? *Journal of Knowledge Management* 20(5), 1004–1024. DOI: 10.1108/JKM-11-2015-0429.

Beise, M. and Rennings, K. (2005) Lead markets and regulation: a framework for analyzing the international diffusion of environmental innovations. *Ecological Economics* 52(1), 5–17. DOI: 10.1016/j.ecolecon.2004.06.007.

Bouchon, F., Hussain, K. and Konar, R. (2017) Event management education and event industry: a case of Malaysia. *Malaysian Online Journal of Educational Management* 3(1), 1–17.

Butler, J. (2008) The compelling "hard case" for "green" hotel development. *Cornell Hospitality Quarterly* 49(3), 234–244. DOI: 10.1177/1938965508322174.

Byrne, S.M., Allen, K.L., Dove, E.R., Watt, F.J. and Nathan, P.R. (2008) The reliability and validity of the dichotomous thinking in eating disorders scale. *Eating Behaviors* 9(2), 154–162. DOI: 10.1016/j.eatbeh.2007.07.002.

Cainelli, G., De Marchi, V. and Grandinetti, R. (2015) Does the development of environmental innovation require different resources? Evidence from Spanish manufacturing firms. *Journal of Cleaner Production* 94, 211–220. DOI: 10.1016/j.jclepro.2015.02.008.

Chang, C.H. (2011) The influence of corporate environmental ethics on competitive advantage: the mediation role of green innovation. *Journal of Business Ethics* 104(3), 361–370. DOI: 10.1007/s10551-011-0914-x.

Chang, C.H. (2018) How to enhance green service and green product innovation performance? The roles of inward and outward capabilities. *Corporate Social Responsibility and Environmental Management* 25(4), 411–425. DOI: 10.1002/csr.1469.

Chang, C.H. and Chen, Y.S. (2013) Green organizational identity and green innovation. *Management Decision* 51(5), 1056–1070. DOI: 10.1108/MD-09-2011-0314.

Chaudhary, S. and Batra, S. (2018) Absorptive capacity and small family firm performance: exploring the mediation processes. *Journal of Knowledge Management* 22(6), 1201–1216. DOI: 10.1108/JKM-01-2017-0047.

Chen, Y.S., Lin, M.J.J. and Chang, C.H. (2009) The positive effects of relationship learning and absorptive capacity on innovation performance and competitive advantage in industrial markets. *Industrial Marketing Management* 38(2), 152–158. DOI: 10.1016/j.indmarman.2008.12.003.

Chen, Y.S., Chang, C.H. and Lin, Y.H. (2014) The determinants of green radical and incremental innovation performance: green shared vision, green absorptive capacity, and green organizational ambidexterity. *Sustainability* 6(11), 7787–7806. DOI: 10.3390/su6117787.

Chen, Y.S., Lin, Y.H., Lin, C.Y. and Chang, C.W. (2015) Enhancement of green absorptive capacity, green dynamic capacities and green service innovation to improve firm performance: an analysis of structural equation modeling (SEM). *Sustainability* 7(11), 15674–15692. DOI: 10.3390/su71115674.
Chong, A.Y.L. and Zhou, L. (2014) Demand chain management: relationships between external antecedents, web-based integration and service innovation performance. *International Journal of Production Economics* 154, 48–58. DOI: 10.1016/j.ijpe.2014.04.005.
Cohen, W.M. and Levinthal, D.A. (1990) Absorptive capacity: a new perspective on learning and innovation. *Administrative Science Quarterly* 35, 128–152. DOI: 10.2307/2393553.
De Marchi, V. (2012) Environmental innovation and R&D cooperation: empirical evidence from Spanish manufacturing firms. *Research Policy* 41(3), 614–623. DOI: 10.1016/j.respol.2011.10.002.
Duan, Y., Wang, W. and Zhou, W. (2020) The multiple mediation effect of absorptive capacity on the organizational slack and innovation performance of high-tech manufacturing firms: evidence from Chinese firms. *International Journal of Production Economics* 229, 107754. DOI: 10.1016/j.ijpe.2020.107754.
Duchek, S. (2013) Capturing absorptive capacity: a critical review and future prospects. *Schmalenbach Business Review* 65(3), 312–329. DOI: 10.1007/BF03396860.
Fukey, L.N. and Issac, S.S. (2014) Connect among green, sustainability and hotel industry: a prospective simulation study. *Energy Conservation* 6(8).
Gabler, C.B., Richey Jr, R.G. and Rapp, A. (2015) Developing an eco-capability through environmental orientation and organizational innovativeness. *Industrial Marketing Management* 45, 151–161. DOI: 10.1016/j.indmarman.2015.02.014.
Hair, J.F., Risher, J.J., Sarstedt, M. and Ringle, C.M. (2019) When to use and how to report the results of PLS-SEM. *European Business Review* 31(1), 2–24. DOI: 10.1108/EBR-11-2018-0203.
Hart, S.L. (1995) A natural-resource-based view of the firm. *Academy of Management Review* 20(4), 986–1014. DOI: 10.5465/amr.1995.9512280033.
Hashim, R., Bock, A.J. and Cooper, S. (2015) The relationship between absorptive capacity and green innovation. *International Journal of Industrial and Manufacturing Engineering* 9(4), 1065–1072.
Henseler, J., Ringle, C.M. and Sarstedt, M. (2015) A new criterion for assessing discriminant validity in variance-based structural equation modeling. *Journal of the Academy of Marketing Science* 43(1), 115–135. DOI: 10.1007/s11747-014-0403-8.
Henseler, J., Hubona, G. and Ray, P.A. (2016) Using PLS path modeling in new technology research: updated guidelines. *Industrial Management & Data Systems* 116(1), 2–20. DOI: 10.1108/IMDS-09-2015-0382.
Homburg, C. and Pflesser, C. (2000) A multiple-layer model of market-oriented organizational culture: measurement issues and performance outcomes. *Journal of Marketing Research* 37(4), 449–462. DOI: 10.1509/jmkr.37.4.449.18786.
Huang, J.W. and Li, Y.H. (2017) Green innovation and performance: the view of organizational capability and social reciprocity. *Journal of Business Ethics* 145(2), 309–324. DOI: 10.1007/s10551-015-2903-y.
Hussain, K., Konar, R. and Ali, F. (2016) Measuring service innovation performance through team culture and knowledge sharing behaviour in hotel services: a PLS approach. *Procedia - Social and Behavioral Sciences* 224, 35–43. DOI: 10.1016/j.sbspro.2016.05.397.
Kasim, A. (2009) Managerial attitudes towards environmental management among small and medium hotels in Kuala Lumpur. *Journal of Sustainable Tourism* 17(6), 709–725. DOI: 10.1080/09669580902928468.
Konar, R., Ali, F. and Hussain, K. (2018) Empowerment in hospitality service leadership: a moderated mediation approach. *Asia-Pacific Journal of Innovation in Hospitality and Tourism* 7(2), 21–38.
Lane, P.J., Salk, J.E. and Lyles, M.A. (2001) Absorptive capacity, learning, and performance in international joint ventures. *Strategic Management Journal* 22(12), 1139–1161. DOI: 10.1002/smj.206.
Lee, K.H. and Min, B. (2015) Green R&D for eco-innovation and its impact on carbon emissions and firm performance. *Journal of Cleaner Production* 108, 534–542. DOI: 10.1016/j.jclepro.2015.05.114.
Lee, K.H., Cin, B.C. and Lee, E.Y. (2016) Environmental responsibility and firm performance: the application of an environmental, social and governance model. *Business Strategy and the Environment* 25(1), 40–53. DOI: 10.1002/bse.1855.
Lichtenthaler, U. (2009) Absorptive capacity, environmental turbulence, and the complementarity of organizational learning processes. *Academy of Management Journal* 52(4), 822–846. DOI: 10.5465/amj.2009.43670902.
Lin, L. (2013) The impact of service innovation on firm performance. *The Service Industries Journal* 33(15–16), 1599–1632. DOI: 10.1080/02642069.2011.638712.

Lin, M.-H., Chen, H. and Liu, kuang-S. (2017) A study of the effects of digital learning on learning motivation and learning outcome. *Eurasia Journal of Mathematics, Science and Technology Education* 13(7), 3553–3564. DOI: 10.12973/eurasia.2017.00744a.

Marzucchi, A. and Montresor, S. (2017) Forms of knowledge and eco-innovation modes: evidence from Spanish manufacturing firms. *Ecological Economics* 131, 208–221. DOI: 10.1016/j.ecolecon.2016.08.032.

Podsakoff, P.M., MacKenzie, S.B., Lee, J.Y. and Podsakoff, N.P. (2003) Common method biases in behavioral research: a critical review of the literature and recommended remedies. *Journal of Applied Psychology* 88(5), 879–903. DOI: 10.1037/0021-9010.88.5.879.

Porter, M. and Reinhardt, F.L. (2007) A strategic approach to climate. *Harvard Business Review* 85(10), 22–26.

Pujari, D. (2006) Eco-innovation and new product development: understanding the influences on market performance. *Technovation* 26(1), 76–85. DOI: 10.1016/j.technovation.2004.07.006.

Pujari, D., Wright, G. and Peattie, K. (2003) Green and competitive: influences on environmental new product development performance. *Journal of Business Research* 56(8), 657–671.

Yusof, Z.B. and Jamaludin, M. (2014) Barriers of Malaysian green hotels and resorts. *Procedia - Social and Behavioral Sciences* 153, 501–509. DOI: 10.1016/j.sbspro.2014.10.083.

Zhang, D., Rong, Z. and Ji, Q. (2019) Green innovation and firm performance: evidence from listed companies in China. *Resources, Conservation and Recycling* 144, 48–55. DOI: 10.1016/j.resconrec.2019.01.023.

Part 3: Communities, Festivals and Events

9 Inclusive Community Development through Events: Lessons from Casual Day with Persons with Disabilities Event in South Africa

Tawanda Makuyana[1]* and Zukiswa Yoliswa Zinhle Nzo[2]
[1]North West University, Potchefstroom, South Africa; [2]National Council of and for Persons with Disabilities, Edenvale, South Africa

Introduction

Giving circles (philanthropy) in the UK and Ireland have self-identified themselves as 'collaboratives' within a social context (Eikenberry and Breeze, 2015). Thus, collaborative giving is explained as groups of donors who collaborate to support individuals, charitable organizations or projects of mutual interest (Eikenberry and Breeze, 2015). In these arrangements, members often conduct collective research on potential beneficiaries and make joint or coordinated decisions about the use of resources (Eikenberry and Breeze, 2015). The 21st century witnessed a transformation from a cross-sector collaboration continuum as 'philanthropic – transactional – integrative' to explain how businesses and non-profit organizations (NPOs) work together. Based on Miller-Stevens and Taylor's (2020) interpretation, this continuum is the philanthropic relationship between businesses and NPOs which fosters relationships where the level of engagement is low and resource commitment is only in the form of a financial transaction and nothing more. Thus, businesses make annual donations in response to requests from NPOs, and the gesture is often considered as value added to the business's marketing strategy (Miller-Stevens and Taylor, 2020). As time has passed, traditional cheque-writing philanthropic relationships have evolved into more sophisticated and complex forms of cross-sector collaboration within the event tourism industry, and disability inclusive community development. This implies that the event industry is a vehicle for people to show their ability to adapt to processes of participating in structured event-groups to solve social and environmental issues, while sharing resources and increasing opportunity towards empowering organizations, staff and previously marginalized or under-represented population groups.

The impacts of institutional structures, which include agency-based collaborations where government organizations dominate, mixed collaborations where both government agencies and citizens are involved, or citizen-based collaborations organized at the grassroots level, have been explored (Diaz-Kope *et al.*, 2015). According to Miller-Stevens and Taylor (2020), the structure of the collaboration is important to consider in that different structures lead to

*Corresponding author: makuyanatawanda@gmail.com

DOI: 10.1079/9781789248685.0009

different processes and outcomes of philanthropic event collaboration among beneficiaries. The ultimate goal of interpersonal collaboration is a cyclical process that leads to respect, and social inclusion for community development.

Casual Day with Persons with Disabilities (Casual Day) is a philanthropic event that embeds multi-stakeholder collaboration within an Afro-centric disability campaign for inclusive societal contexts. Thus, the organizers of Casual Day foster the belief that events need to achieve emotional satisfaction, enjoyment satisfaction and experiences satisfaction among the targeted audiences (Wood and Moss, 2015). Within this understanding, Casual Day adheres to the generic event processes, namely research, designing, planning, coordination and evaluation (De Geus *et al.*, 2016). Based on Nguyen *et al.*'s (2019) interpretation of best practices in event management, Casual Day's process incorporates factors such as time, finance, technology and human resources as critical. This poses pressure and emphasis on a quality experience in philanthropic events and festivals (De Geus *et al.*, 2016), economic and developmental impacts and visitor motivations (Gursoy *et al.*, 2004; Getz, 2005), operationalizing and quantifying the experiences, regardless of a lack of consensus on the definition of the essence of experiences within philanthropic event management (Walls *et al.*, 2011).

There is a dearth of literature connecting philanthropic events, accessibility/inclusion, and opportunities to empower disabled people for collaborative partnerships and participation towards disability-inclusive community development (Miller-Stevens and Taylor, 2020). Yet there are intuitive connections for sustainable inclusive community development, which remains unaddressed by opinions and/or research-based opinions from the tourism event industry, academics and the disability sector. While adopting descriptive-narrative research, reality has emerged from people's empirical experiences for this work to provide insights into various views associated with the future of the philanthropic events tourism industry and disability inclusive community development. Thus, as the existing literature does not provide a detailed narrative on the connection, this chapter takes inspiration from qualitative data existing in the secondary literature and empirical experiences of Casual Day with Persons with Disabilities in South Africa.

Conceptualizing Casual Day with Persons with Disabilities

Events have been defined as a one-time or infrequently occurring event of limited duration that provides the consumer with a leisure and social opportunity beyond their everyday experience (Van Vliet, 2012). A festival is a gathering of a relatively large crowd in a specific public place for a delineated period, during which visitors (the audience) are offered a unique experience (planned and organized with a specific purpose), including transformation and play elements, making it possible for visitors to behave and feel different (emotional satisfaction, experience and enjoyment) from how they feel in their daily lives (Van Vliet, 2012, p. 20). Philanthropy is goodwill to a person through active effort to promote human welfare through: (i) a charitable/benevolent act or gift; or (ii) an organization distributing or being supported by humanitarian funds (Merriam-Webster Online, 2008). One therefore regards the definition of a philanthropic event and festival as a combined field, which can be defined as a gathering of a crowd in a public space (physically and/or virtually) for a delineated period in pursuit of transforming value addition to others, while providing the consumer (the audience) and beneficiaries with a leisure and socio-economic opportunity beyond everyday experience that contributes to community development (Sulek, 2009).

As noted above, Casual Day is an avenue for advocacy on disability inclusion that upholds the pursuit of education, awareness and empowerment for community growth, as per the United Nations Convention on the Rights of Persons with Disabilities (UNCRPD) (United Nations Department of Economic and Social Affairs, 2006). Thus, Article 8 of UNCRPD (2006) emphasizes parties' commitment to raising awareness in order to foster respect for the rights and dignity of disabled people and to counter disability discrimination throughout society, thereby combatting stereotypes, prejudices and harmful behavioural practices to individuals with impairments. Therefore, Casual Day is

Fig. 9.1. Conceptual framework for inclusive community development for the example of Casual Day with Persons with Disabilities in South Africa.

illustrated by a conceptual framework in Fig. 9.1 as inclusive interactive experiences between the audience and the event environment (physical or virtual) that foster various levels of engagement of multiple experiential elements and outputs like satisfaction, emotions, behaviours, cognition, memories and learning, that can happen and/or create the desired impact at any point in the event journey (successful celebration of diversity).

Based on Fig. 9.1, Casual Day contributes to disability-inclusive community development by the inclusion of disabled women, men, girls and boys in their local communities (Barman, 2017). This is because Casual Day is an inclusive philanthropic fundraising and flagship event for the advocacy and awareness campaign that advances disability awareness and education. Within this context, the terms 'disability' and 'impairments' are used interchangeably, regardless of the fact that 'disability' refers to the physical and non-physical barriers in the environment (Disabled World, 2016), while 'impairment' refers to individuals' deformation of limb or body part (UNCRPD, 2006). In this chapter, being a disabled person refers to participation restriction resulting from the interaction between persons with impairments and attitudinal and environmental barriers that hinder their full and effective participation in society on an equal basis with others (Makuyana, 2020). To this effect, the chapter prefers usage of the terms 'disabled people and persons/individuals with impairments', rather than 'persons with disability', unless when referring to the name of the project 'Casual Day with Persons with Disabilities' (Casual Day) or names of organizations.

Origin of Casual Day with Persons with Disabilities in South Africa

The National Council of & for Persons with Disabilities (NCPD) is an organization that was founded 80 years ago. It serves people with impairments and has a countrywide footprint in South Africa using the Associations for Persons with Disabilities (APDs). Casual Day emerged as a concept created by NCPD in 1994. It was operationalized in 1995 as a mere invitation to all South Africans to dress differently (diversity). Thus, an opportunity for disabled people to have fun, just like on a 'fun day event' while raising awareness and visibility of disabled people in local communities (*Casual Day Operating Manual*; NCPD, 2019). This initiative furthered disability inclusion efforts that began in the 1970s in South Africa. Casual Day made society realize the need for social inclusion.

Prior to the coronavirus disease 2019 (COVID-19) pandemic, the Casual Day concept was a day when one dressed in 'civvies' (i.e. civilian clothes as opposed to a uniform) on the first Friday of September in the workplace and in schools. Thus, 'civvies' was a gesture of bringing diversity of appearance/presentation among the team in an organization, yet the individuals work as one organ towards the pursuit of one organizational goal. This implied that disabled people may not look like non-disabled in appearance, but they are an important part of society. Additionally, September is Tourism month, whereby NCPD and Casual Day advocate for accessible tourism in South Africa (Makuyana *et al.*, 2020). Participation was achieved by a donation of R5 in buying a Casual Day physical sticker. The value increased to R10 on the celebration of Casual Day's tenth anniversary. In 2020, lockdown restrictions induced by the COVID-19 pandemic meant that Casual Day was celebrated on 3 December – International Day of Persons with Disabilities. This coincided with the day on which the World Tourism Organization (UNWTO) unveiled a research report that regarded accessible/inclusive tourism as a game changer for sustainable tourism recovery in the period post-COVID-19 pandemic and lockdown. During the pandemic participants paid R20 to buy an official Casual Day digital sticker as a mechanism to adhere to social distancing and working/celebrating using virtual platforms, while being at home. In COVID-19 times, the digital sticker not only helps raise funds for national beneficiaries, but also boosts awareness about disability throughout South Africa, while maintaining the 'civvies' context.

As time passed, the disability discourse evolved and integrated international and domestic legislative frameworks that acknowledge inclusion of disabled people. Thereby this accelerated the need to empower and enhance the participation of disabled people throughout the event processes beyond just receiving the proceeds of charity (Makuyana, 2020). The past decade has witnessed a transformation in the objectives of Casual Day, from just being a fundraiser and advocacy for disabled people to have fun and interact, to giving disabled people opportunities to earn some money, while rendering services to Casual Day. Thus, Casual Day became an annual event that had the preparation process, execution and the post-event that involves, empowers and enhances active participation (visibility) of disabled people and persons/individuals with impairments, so they are not just the beneficiaries of charity. Thus, disability issues are continuously addressed by this event as an evolving event for the past 26 years (NCPD, 2019). Hence, Casual Day became an instrument for disability advocacy, awareness, education, participation, support and empowerment of disabled people for the creation of an inclusive society. This has been pursued through collaborative relationships and/or partnerships with non-disabled counterparts (community membership), disabled youth and women, corporates, disabled people organizations (NPOs) and community-based organizations (non-governmental organizations, NGOs), schools for disabled learners, mainstream schools and universities. The private tourism-related companies include, but are not limited to, hotel chains and Disney Land South Africa (as main sponsor for 2 years in a row: 2020 and 2021), Edcon group (as official partner for 9 years), government and quasi-government authorities/agents.

Casual Day Framework Model

Like any other philanthropic event, the Casual Day project has 14 National Disabled Peoples Organizations (NDPOs) and NCPD, who share 60% of the financial resources that are raised (see Fig. 9.2). Nonetheless, the NDPOs are encouraged to take responsibility for being the agent selling the Casual Day merchandise and retain their 40% portion of the sales value for their own use (see Fig. 9.2). In this context, the portion of NCPD is divided towards its affiliated nine provincial APDs and over 600 community-based organizations concerned with disabled people, as their mandated beneficiaries. This is because Casual Day runs as a continuous event (project) throughout the year, in spite of having a day of observation of 'diversity' as a metaphor set by 'civvies'. Therefore, Casual Day showcases: (i) the ability to enhance disability awareness and education to society; (ii) the outcomes (positive and adverse impacts) to local community

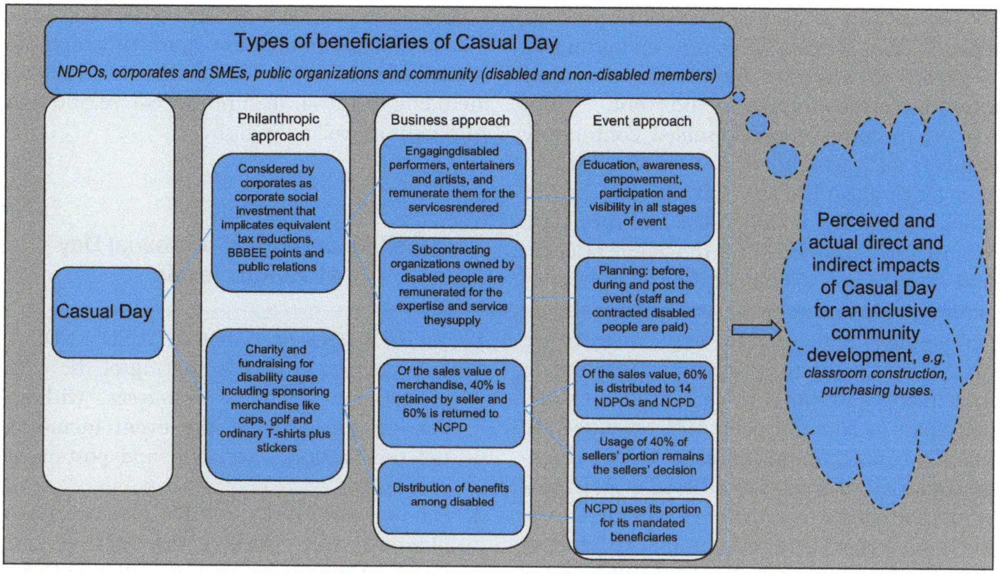

Fig. 9.2. Casual Day framework model: integrated philanthropic, business and event approach for inclusive community development. BBBEE, Broad-based Black Economic Empowerment; NCPD, National Council of & for Persons with Disabilities; NDPOs, National Disabled Peoples Organizations; SMEs, small and medium-sized enterprises.

development of an inclusive event and festival; (iii) the perceived lessons learnt in the planning process as well as business intelligence gained over the past 26 years; and (iv) the perceived leveraging and/or success factors towards disability-inclusive community development.

Based on the framework model (shown in Fig. 9.2), Casual Day has categories of sponsorship and collaborative partnerships which come with terms and conditions such as: (i) making sure that the sponsor's brand is visible; (ii) corporate social investment that would enhance tax reductions for the donations and earn Broad-based Black Economic Empowerment (BBBEE) scorecard-points as part of meaningful engagement within the South African transformation agenda that is rooted in legislative frameworks like the Broad-based Black Economic Empowerment Act of 2013 (South African Government, nd); and (iii) their funding does not cover Casual Day's staff salaries. On the other hand, the public sector is encouraged to participate as a means to pursue the National Development Plan, vision 2030 (South African Government, 2011) and disability quota. These include government agents such as the South Africa Revenue Authority, South Africa Correctional Services and the Police as active participants.

Based on the framework model, 60% of the sales value from Casual Day merchandise and stickers is given back to NCPD and 40% is retained by the private- and public-sector participants who sell the merchandise (see Fig. 9.2). However, each of the organizations who sell the merchandise has the choice of using it for their own needs or donating it for use in Casual Day in pursuit of some of the terms highlighted in the previous paragraph (see also Fig. 9.2). Thus, sponsors and collaborators incur the cost of producing the merchandise and the associated sales and distribution (logistics). This implies that Casual Day operationalize aspects of conventional business approaches through outsourcing services and expertise from disabled entrepreneurs, artists, performers, marketers, among other needs, as paid subcontractors for the event. This contributes towards promotion, ensuring visibility of otherwise traditionally hidden professionals, and supporting their empowerment instead of getting such services from voluntary non-disabled counterparts.

The volunteering disabled peers are given a stipend. The services are not compromised in quality, as witnessed by the past record of success since the inception of the event.

On the other hand, disabled community members such as schools and universities that cater for disabled learners have been points of selling Casual Day merchandise and they have retained their 40% portion of the selling prices to fund their own development. The non-disabled community members who participate in selling merchandise and stickers (e.g. mainstream schools, youth groups, churches, government departments and South African citizens at large) participate by donating their 40% revenue from selling the merchandise to Casual Day. As highlighted above, all revenue gathered from Casual Day is then shared among the 14 NDPOs. Casual Day has managed to raise a cumulative figure of above R404 million since 1995. Auditing has been conducted after the events each year, prior to the distribution of the 60% to the other 14 NDPOs. The post-event marks the inception of the planning process, which includes relationship management for its partners, feedback and monetary and non-monetary impact assessment reports each year and the related costs are not incurred by partners and collaborators.

Casual Day as a Vehicle for Disability Education and Awareness in Society

Communities who are exposed to Casual Day exhibit attitudes with greater accommodation and learnt behaviours with reduced stereotyping towards disabled people and persons/individuals with impairments. This is evident by the ever-increasing numbers of new participants. On the one hand, it seems that Casual Day is contributing to the process of unlearning previously learnt discriminative behaviours and bias against disabled people within societal normalcy. On the other hand, Casual Day has room to improve its approaches to educating about disability from a more social model perspective.

Disability awareness and education is broadcast using radio and television stations, as well as social media such as Twitter and Facebook in addition to NCPD and Casual Day websites. Casual Day uses community radio stations to reach local communities using their own languages. In addition, marketing material is available in various South African languages including seXhosa, seSotho, sePedi, seZulu and Afrikaans as well as English.

Positive outcomes of Casual Day at various levels

1. *The 14 NDPOs in relation to NCPD* – NDPOs only partake in benefit sharing of the 60% of gathered financial resources, without bearing the costs of the event including the preparation, execution and post-event processes.
2. *Nine provincial APDs* – NCPD uses its portion obtained from sharing the 60% of the financial resources to fund all provincial APDs and local-community disabled people organizations as part of their mandate. In addition, NCPD encourages the APDs and local-community-based organizations to partake in selling the merchandise for them to retain 40% for their own organizational use.
3. *District/local community (disabled people, non-disabled counterparts within community development)* – Of the sales value of merchandise, 40% is retained for community development, for example construction of classrooms at a school that caters for disabled learners, purchase of a bus, or use of the Casual Day golf shirts as part of uniforms. The participation has no transfer of risk of losses or costs because NCPD organizes the distribution (pays for all courier costs) of the merchandise to the selling agents (participants) and the retained portion is more of a funding support to them.

Thus, the impact of the financial resources raised by Casual Day is not limited to donations channelled to individuals with impairments, but also contributes as follows:

- Schools in Limpopo Province and North West Province, which include but are not limited to Tshilidzini Special School (TSS) for blind, deaf and physically disabled students, used financial resources from Casual

Day to buy assets such as bakkies (small trucks).

- Casual Day financial resources have been used to drill boreholes for gardening, for assistive devices such as computers and Braille machines, and to pay for staff working at homes and hostels for disabled people.
- Workshops for disabled people used financial resources from Casual Day to build a classroom for woodwork (an example of vocational training) for disabled people.
- Casual Day supports the disability units of 11 universities in South Africa.

The Casual Day staff complement is comprised of disabled personnel as well as contractors. These people provide services and expertise that are remunerated. Examples of such people include an entertainment manager, a cameraman, motivational speakers, entertainers (performers and singers), entrepreneurs who sell merchandise, interpreters and ambassadors. Most of these people are volunteers who receive a stipend to retain their services.

The impact of Casual Day is being felt in the societies of Soweto and Nelson Mandela Bay where there is a breakdown of stereotypes of disabled people. In 2020, Casual Day observation incorporated a virtual-based youth concert organized by 2J (a disabled entertainer and artist with a video/digital company). In 2021, Casual Day was celebrated in September, and evolved towards having a build-up of online talent show events as a means to reinforce a continuous change of narrative. This is called the Casual Friday-platform that showcases disabled artists, media characters and entertainers. The platform was organized again by 2J.

Perceived Lessons Learnt From the Casual Day Event over the Past 26 Years

The Casual Day team and NCPD have learnt lessons as part of the learning curve as presented in Table 9.1. The perceived lessons learnt include:

1. There is a need for a non-segmented audience, that is an audience that includes everyone, and is not just limited to civil servants and certain sectors of society such as teachers. This can reduce the adverse impact of incidents such as nationwide strikes, for example the teachers' strike in 2009.
2. Usage of stickers has been regarded as outdated by the feedback from participants and also observed by the decline of sales, hence the adoption of digital stickers during COVID-19. However, the digital stickers are difficult for some people to acquire if they do not have easy access to computers (e.g. some of the rural population). This has posed more challenges for the middlemen in the distribution chain of the merchandise.
3. Casual Day lacks professional attention on social media platforms like Twitter and Facebook. This impacts the knowledge and value of social media to the event and the effect that influencers can have in converting adverts into sales. There is a need to expand media coverage.
4. Disabled people failed to utilize marketing, advertising and public relations (PR) to generate more sales. Hence, Casual Day needs to develop a capacity for upselling.
5. There is competition from other types of events, and the model can be easily copied.
6. Casual Day has spin-offs, hence if it is assessed only from returns on investment, non-quantifiable impacts would not be brought to the surface, but this can be established when the social return on investment is also assessed. This is important as much of the impact is achieved indirectly by a ripple effect on members of society, so that both discriminatory attitudes and behaviours are unlearnt and inclusion is learnt through disability awareness – education as embedded in empowerment for socio-economic participation.
7. Administration fees are met by the NCPD, which includes but is not limited to salaries for the direct staff who constitute the Casual Day team. However, salaries for people like the CEO and Finance Manager are not considered under the Casual Day expenses

Table 9.1. Perceived lessons learnt in the past 26 years of the Casual Day annual event.[a]

Stages in planning the event	Planning processes and business intelligence (aptitude/acumen) for the philanthropic event	Practice	Lessons learnt
Research (preliminary planning and approval)	Entertainment	Participation, involvement of disabled performers, entertainers and artists	Disabled entertainers, artists and performers can share stage with non-disabled peers
	Facilities, equipment and supplies	Provided by disabled entrepreneurs as paid services	Supporting and empowering disabled people is vital for their economic activeness
	Safety and security	Provided by disabled entrepreneurs	The service provider's team is composed of both disabled and non-disabled people
Designing	Budgeting	Event project budget	There is need to bring self-sustenance of the project throughout the year, as this caters for Casual Day expenditure like structure and management
	Venue	Venue was only virtual in 2020	Virtual venue worked for both NCPD and for individual participating organizations
	Vendor scouting	Engages disabled people and non-disabled when deemed necessary	Involvement of disabled entrepreneurs
	Volunteering and recruiting volunteers	Both disabled and non-disabled people volunteer according to their capacity to contribute to the event	Disabled people are usually paid a stipend to support their cash streams
	Event design	Casual Day team's role is to lead the team made up of disabled and non-disabled event designers/organizers	Disabled event designers/organizers work with non-disabled peers using a teamwork approach
	Event logistics	Casual Day team's responsibility	Disabled and non-disabled colleagues complement each other in event logistics
Event planning (programme planning)	Event coordination	Casual Day team's responsibility	Disabled and non-disabled colleagues complement each other in event coordination

Continued

Table 9.1. Continued

Stages in planning the event	Planning processes and business intelligence (aptitude/acumen) for the philanthropic event	Practice	Lessons learnt
	Decoration	Casual Day team's responsibility	Disabled and non-disabled colleagues complement each other in event decoration
	Signage and banners	Engage a disabled entrepreneur as paid services	Enhancing participation: involving and empowering disabled entrepreneurs can make society see that having an impairment does not necessarily make one able or unable
	Membership and database management	Stakeholder relationship and database management are done by Casual Day team	Business (especially marketing and client relationship management) principles apply
	Legal issues	NCPD has network that supports with legal expertise	Donation of legal expertise (human resources) to complement financial giving
	Sponsors	Casual Day has a blend of corporates (private sector), SMEs, individuals and government agents that have established a collective donation for the past 26 years	Collective/collaborative philanthropy can yield different impacts to different participants. However, the common objective in this is to further disability inclusion as well as empowering disabled counterparts
Coordination	Transportation	NCPD provides such services	Casual Day has no capital assets; it relies on NCPD. However, this should be changed for self-sustenance as an autonomous annual event project
	Talent selection and recruitment	Casual Day team's responsibility	Collaborative engagement with disabled recruitment networks/agents yields results
	Promotion	Paid services are provided by disabled entrepreneurs	These actions that support disabled entrepreneurs can also be good and active in economic transaction and service provision as part of social normalcy

Continued

Table 9.1. Continued

Stages in planning the event	Planning processes and business intelligence (aptitude/acumen) for the philanthropic event	Practice	Lessons learnt
	Publicity	Paid services are provided by disabled entrepreneurs	These actions that support disabled entrepreneurs can also be good and active in economic transaction and service provision as part of social normalcy
	Printed collateral[b]	Paid services are provided by disabled entrepreneurs	These actions that support disabled entrepreneurs can also be good and active in economic transaction and service provision as part of social normalcy
	Event execution (tools and methods implemented)	Joint work of disabled and non-disabled stakeholders	Inclusive approaches yield better, especially in breaking down stereotyping of disabled people and non-disabled counterparts
	Target group	Countrywide (non-segmented audience)	Everyone who can be reached by the Casual Day campaign can benefit in one way or the other towards inclusion of disability and disabled people
	Communication with audience:		
	Internal External Online marketing Other marketing tools	Television, Internet-based, radio, community radio (using local languages), social media (Twitter, Facebook), word of mouth, merchandise (e.g. stickers, T-shirts, golf shirts, caps, hats, wristbands, masks with Casual Day disability awareness messages)	People can hear the message and learn, which can bring unlearning of discriminatory attitudes and behaviours to become accommodative, though it is a process. Thus, observed feedback shows communities that participate in Casual Day have shown more accommodating attitudes and behaviours
	Presentation of product	The product is disability inclusion that is packaged in different forms as well as made tangible in a variety of ways	There is need to bring perceived value in the way the message is packed among all stakeholders
	Event set up	Casual Day team coordinates; however, there is room for tailored approach among different participating stakeholders	Giving stakeholders room to be innovative enhances innovation as well as contextualizing the campaign for diversity and inclusion

Continued

Table 9.1. Continued

Stages in planning the event	Planning processes and business intelligence (aptitude/acumen) for the philanthropic event	Practice	Lessons learnt
	Event duration	It is officially throughout the year, however, with a day of observation (namely 3 September, except in 2020 it was on 3 December)	It has adopted characteristics of a continuous event that is celebrated/observed annually on a certain date (virtually and physically)
	Event take down	It used to be physical until 2020 when, due to the global pandemic, a virtual approach was adopted	Contextualized innovation and business intelligence are critical for survival of Casual Day
Evaluation	Event evaluation comprising analysis of the process and product: (i) goals; (ii) overall reflection of event; and (iii) professional development	Audits are done after the event roll-out, and feedback is sourced from the stakeholders, which would be set in a report to be discussed by the Casual Day team as part of reconciliation and reflection for professional development	A continuous monitoring and evaluation of the event for continuous improvement is vital
Time management	Duration and time utilization	Targets and deadlines are set and observed	Benchmarking with previous successes is significant in informing a starting point towards future improvement in time management
Financial	Resource mobilization	Sponsorship and NCPD support for Casual Day	A triangulation of collaborative funding, events management and business approaches are the bedrock of Casual Day's success story
	Optimal utilization	Casual Day model and budget guide the event processes	The Casual Day model needs remodelling to address a balance of social model issues (e.g. the value contributed by 14 NDPOs' participation in the event)
	Management	Casual Day team's responsibility	A joint management approach is crucial, especially with the other 14 NDPOs
	Accounting and accountability	Casual Day and NCPD provide accounts and are accountable	A joint approach that engages other stakeholders is needed; however, NCPD can be the lead accounting entity

Continued

Table 9.1. Continued

Stages in planning the event	Planning processes and business intelligence (aptitude/acumen) for the philanthropic event	Practice	Lessons learnt
Adoption of technological efficiency	Risk management and innovation	The current approach takes away financial risk from the middleman who sells the merchandise	Innovation is needed, especially that can bring transformative value of the merchandising approach among other aspects in the Casual Day model to be contextual as well as being updated as enhancing and reinforcing perceived value among stakeholders
Human resources	Expertise/competencies and knowledgeability	Casual Day depends on the initiative and other marketing skills of participants	There is a need to upskill and/or facilitate capacity development of participants, e.g. through entrepreneurial skills development on sales (ability to convert social media influence, advertising and marketing into sales), digital skills and Casual Day product innovations to address socio-economic preferences and tastes. Also a need to make the digital stickers available to all participants so they can touch and feel them in such a way that Casual Day retains its traditional real substance that it had before the COVID-19 pandemic

[a]NCPD, National Council of & for Persons with Disabilities; NDPO, National Disabled Peoples Organizations; SMEs, small and medium-sized enterprises.
[b]Printed collateral refers to any printed marketing material used to promote, campaign and market the Casual Day event.

yet they are directly involved in all processes and planning for Casual Day.

8. The NCPD and Casual Day policy do not explicitly highlight financial considerations, charges and/or costs related to Casual Day as deducted from the revenues generated by Casual Day.
9. NCPD and Casual Day policy do not enable self-sustainability of the Casual Day project by putting aside resources for the next Casual Day event. Thus, each year's event starts from zero resources, thereby the main sponsor needs to provide financial support for the event budget each year.
10. The NCPD and Casual Day do not explicitly highlight the assets for the Casual Day as being independent from NCPD, so the financial statements for Casual Day do not reflect the capital assets in spite of usage of NCPD's assets for the Casual Day project. Hence, Casual Day is not a standalone project of NCPD. This situation implies that overheads are high and this compromises the viability and worthiness of the Casual Day project. Hence, the nature of the organization (NCPD-Casual Day) cannot be evaluated from a cost equation. However, socio-economic impacts should be taken into consideration in economic evaluation of the event project.
11. Casual Day brand equity has been measured among the targeted audience/market for the Casual Day and it was found that the commercial value derived from consumer perceptions of the brand name upheld a high socio-economic impact in terms of philanthropic performance.
12. Selling of merchandise through the middlemen is being strained as the individual end user has more options of accessing the merchandise within the distribution channel. In addition, courier costs for the unsold merchandise are added to the costs of Casual Day and NCPD, yet the merchandise cannot be used in the next event because they are designed according to themes, hence this creates more risk for the NCPD.

Recommendations for continuous improvement of Casual Day are presented in Table 9.2. These recommendations arise from primary data collected from key informants who were interviewed using Google Meet, a video-communication service, in June 2021. Interviewees included the Head of Marketing and Communication of one of the main sponsors Alpha Pharmacies, the National Director (CEO) of NCPD, four Casual Day team members, and three performers (artists). Other interviewees were beneficiaries and participants of Casual Day, including five leaders of special schools or care centres from Limpopo and North West provinces, two representatives of APDs in Free State and Western Cape provinces, members of four workshops in the above-mentioned provinces, four disability units from universities in South Africa and 14 NDPOs. A semi-structured interview guide was used and the interview discussion took between 30 minutes and

Table 9.2. Perceived areas for improvement for the Casual Day annual event project. From primary data collected through interviews with key informants.

Area to be improved	Role in Casual Day	Mechanism to enhance the change	Operationalize the change
Upselling	Middlemen (selling agents)	Bringing these skills into programmes offered by NCPD and their capacity development networks	Enhancing 'sitting-next-to-Nellie' approach (experiential/on-the-job learning) as augmented by formal training on events management for community development
Sale skills	Middlemen (selling agents)		
Innovativeness[a]	Event planners (organizers)		
Perceived value	End-user (beneficiaries/ target audience) and selling agents		

[a]Innovation is required in all areas of the project but particularly regarding the need to make digital stickers available to everyone, even those without easy access to computers.

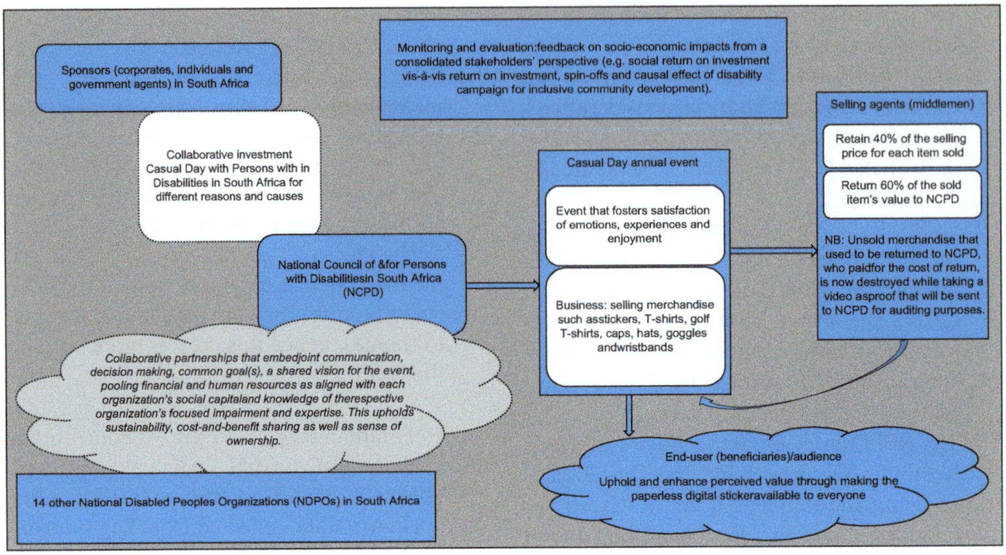

Fig. 9.3. Framework for inclusive community development using philanthropic events, using Casual Day as an example.

2 hours. The aim was to gather opinions on the interviewee's participation/support for Casual Day's objective to further inclusion of disability and disabled people for community development. The interview discussion entailed: (i) their motivation behind their support/participation; (ii) their role and involvement; and (iii) the perceived impact towards inclusion of disabled children and adults and community development.

Based on Table 9.2, Casual Day should remould its model and/or approach in order to bring a sense of ownership to the other 14 NDPOs as well as provincial APDs. This suggests there is a need to evolve Casual Day towards self-sustaining approaches in event financial resource mobilization and optimizing the use thereof, while creating reserves and/or capital. It is vital to bring innovation to make the digital stickers available to everyone.

In addition, the primary data provided the information that enabled the development of the framework for inclusive community development using philanthropic events as presented in Fig. 9.3. Figure 9.3 incorporates recommendations that foster sustainable inclusive community development using philanthropic events that embed business approaches such as Casual Day.

Perceived Leveraging/Success Factors of Casual Day

The perceived leveraging factors of Casual Day include, but are not limited to, the following:

1. Incorporation of disabled people at the forefront of the Casual Day operationalizes empowerment, involvement and participation of disabled persons.
2. Awareness and education covers a wider audience for wider reach in breakdown of stereotyping among communities that have knowledge of Casual Day as compared with communities that are not exposed to disability knowledge and Casual Day.
3. There are opportunities for disabled people to interact while being given attention in the media. This upholds visibility and a platform to air the self-representative voice of disabled people.
4. Monetary value and benefits that directly and indirectly reach disabled people through the involvement and participation in the Casual Day project and programmes are empowering disabled people.
5. There has been growth and innovation since the inception of Casual Day in 1995.

6. There is visible and direct inclusive community development through schools for disabled learners that use their 40% to build classrooms, buses, boreholes as well as co-opting the golf shirts of Casual Day to be part of their uniform as a mechanism to further disability education and awareness in the communities.
7. The Casual Day themes are motivated by consultation with trendsetters, sponsors and disabled people's ideas as aligned with inclusivity, value-based diversity and in the context with what is popular at the time.

The above primary data is summarized as a leveraging matrix for Casual Day as shown in Table 9.3.

Conclusion

This chapter presents a framework for a philanthropic event, Casual Day with Persons with Disabilities, that increases accessibility and opportunities for disabled people to participate in community development in South Africa. The chapter nurtures a narrative of enhancing communities' understanding and relationships with disabled people as part of an inclusive society. Specifically the chapter describes the feasibility of developing an inclusive and sustainable event that enhances disability awareness and education to society and the events tourism industry.

Table 9.3. Leveraging implementation matrix for inclusive community development using philanthropic events, using Casual Day as an example.

Business intelligence in events	Casual Day variables[a,b]								
	Time management	Human resources/ expertise	Government legislative frameworks	Corporate/social investment	Individual ethical/moral diversity	Technological efficiencies and ability to adapt	Community learnt behaviours and attitudes	Social capital	Institutional structure
Research	X	X	X	X	X	X	X	X	X
Design	X	X	X	X	X	X	X	X	X
Event planning	X	X	X	X	X	X	X	X	X
Coordination	X	X	X	X	X	X	X	X	X
Evaluation	X	X	X	X	X	X	X	X	X

[a]Casual Day variables are those that are integrated into the planning process.

[b]Outcomes: (i) building and nurturing education and awareness on inclusion of disability and disabled persons; (ii) demonstration of knowledge, appreciation and eagerness to apply learnt behaviour on diversity and disability inclusion within societal systems; and (iii) efforts towards operationalizing inclusive fundraising and awareness campaign framework for inclusive community development by Casual Day with persons with disabilities' stakeholders in South Africa.

References

Barman, E. (2017) The social bases of philanthropy. *Annual Review of Sociology* 43, 271–290.

De Geus, S., Richards, G. and Toepoel, V. (2016) Conceptualisation and operationalisation of event and festival experiences: creation of an event experience scale. *Scandinavian Journal of Hospitality and Tourism* 16(3), 274–296. DOI: 10.1080/15022250.2015.1101933.

Diaz-Kope, L., Miller-Stevens, K. and Morris, J.C. (2015) Collaboration processes and institutional structure: reexamining the black box. *International Journal of Public Administration* 38(9), 607–615. DOI: 10.1080/01900692.2014.949755.

Disabled World (2016) Available at: www.disabled-world.com (accessed 20 January 2021).

Eikenberry, A.M. and Breeze, B. (2015) Growing philanthropy through collaboration: the landscape of giving circles in the United Kingdom and Ireland. *Voluntary Sector Review* 6(1), 41–59. DOI: 10.1332/204080515X14241771107299.

Getz, D. (2005) *Event Management and Event Tourism*, 2nd edn. Cognizant Communication Corporation, New York.

Gursoy, D., Kim, K. and Uysal, M. (2004) Perceived impacts of festivals and special events by organizers: an extension and validation. *Tourism Management* 25(2), 171–181. DOI: 10.1016/S0261-5177(03)00092-X.

Makuyana, T. (2020) A model for accessible tourism human capital development for the physical impaired persons in South Africa. PhD thesis. North West University, Potchefstroom, South Africa.

Makuyana, T., Du Plessis, E. and Chikuta, O. (2020) Profiling disabled people as unrealised potentially lucrative tourism market segment in South Africa. *African Journal of Hospitality, Tourism and Leisure* 9(5), 876–891.

Merriam-Webster Online (2008) Available at: https://www.merriam-webster.com/ (accessed 13 July 2021).

Miller-Stevens, K. and Taylor, J.A. (2020) Philanthropic collaboration: a conceptual framework for giving circles. *Public Integrity* 22(6), 1–15. DOI: 10.1080/10999922.2020.1719808.

NCPD (National Council of & for Persons with Disabilities) (2019) *Casual Day Operating Manual: for Participating Beneficiaries*. NCPD, Johannesburg, South Africa.

Nguyen, T., Lee, K., Chung, N. and Koo, C. (2019) The way of generation Y enjoying jazz festival: a case of the Korea (Jarasum) music festival. *Asia Pacific Journal of Tourism Research* 25(1), 52–63.

South African Government (n.d.) Republic of South Africa, Broad-based Black Economic Empowerment Act 53 of 2003, amendment Act 46 of 2013. Available at: www.gov.za/documents (accessed 29 August 2021).

South African Government (2011) Republic of South Africa, National Development Plan, vision 2030. Available at: https://www.gov.za/issues/national-development-plan-2030 (accessed 29 August 2021).

Sulek, M. (2009) On the modern meaning of philanthropy. *Nonprofit and Voluntary Sector Quarterly* 39(2), 193–212. DOI: 10.1177/0899764009333052.

United Nations Department of Economic and Social Affairs (2006) United Nations Convention on the Rights of Persons with Disabilities (UNCRPD). A final report of the Ad Hoc Committee on A Comprehensive and Integral International Convention on the Protection, and Promotion of the Rights and Dignity of Persons with Disabilities [A/61/611.] New York: United Nations.

Van Vliet, H. (2012) Festivals: eenintroductie. In: Van Vliet, H. (ed.) *Festival Beleving. De Waarde van Publieksevenementen*. Hogeschool, Utrecht, the Netherlands, pp. 13–31.

Walls, A.R., Okumus, F., Wang, Y. and Kwun, D.J.-W. (2011) An epistemological view of consumer experiences. *International Journal of Hospitality Management* 30(1), 10–21. DOI: 10.1016/j.ijhm.2010.03.008.

Wood, E.H. and Moss, J. (2015) Capturing emotions: experience sampling at live music events. *Arts and the Market* 5(1), 45–72. DOI: 10.1108/AM-02-2013-0002.

10 Boom Festival – a Study about the Evolution of Local Inhabitants' Perceptions between 2009 and 2021

Sara Morgado Nunes* and Ana Rita Garcia
Polytechnic Institute of Castelo Branco, Castelo Branco, Portugal

Introduction

Every 2 years, Idanha-a-Nova experiences an extraordinary moment. In fact, a semi-rural community, guaranteeing ancestral traditions, possessing a rich heritage, finds itself 'invaded' by a group of visitors who, certainly, do not leave the locals indifferent. Young people (of age and spirit) arrive from all over the world, whom we identify as heterodox in relation to the common. They dress in an unprecedented, colourful way, approaching that of the hippy communities of the 1960s. The defence of the environment, a life close to nature, the disregard for rules, anarchism lived in a full way from food to the relationship between people, are surprising; the population's comments oscillate between laughter, rejection and the economic advantage that results from the appearance, in August on the first full moon, of people who enrich the human cultural mosaic. It is known that tourism is an engine for the development of territories. The demand and supply of cultural events, namely festivals, have grown throughout Portugal, renovating and revitalizing places on both an economic and a landscape level, while preserving the cultural and historical heritage (Ribeiro *et al.*, 2005).

With this investigation, we wanted to know the similarities and differences in the perception of residents in Idanha-a-Nova, regarding the Boom Festival, between the years 2009 and 2021. The questions are:

- Does the local population distance themselves emotionally from the Boom Festival?
- Will they consider the event to be positive in economic terms and in terms of promoting the venue?
- Do they point out negative effects associated with this type of event?

Answering these questions is our goal. Has the perception of the people from Idanha-a-Nova changed over the course of 12 years? This research begins with a review of the literature on music and dance festivals, followed by a description of Idanha-a-Nova. Then, the Boom Festival is explored, reporting the changes it has been promoting over the years. Finally, the results of surveys carried out in 2009 and 2021 among the population living in Idanha-a-Nova are presented, to understand the differences and similarities in the perception of the inhabitants.

*Corresponding author: sara@ipcb.pt

DOI: 10.1079/9781789248685.0010

Summer Music and Dance Festivals

Summer music festivals are 'formal periods or programs of enjoyment activities, entertainment, or events that have a festive character and that publicly celebrate some concept, event or fact' (Janiskee, 1980, p. 97). The duration is short, they occur in a generally limited space; they have well-defined objectives, and tend to have an extensive musical programme (Abreu, 2004). Rolfe says that:

> Some festivals focus only on one form of art. Others include activities in many art forms. In addition to almost all genres of music, there are festivals of dance, drama, cinema, literature, poetry, etc. ... and countless innovative combinations of art forms.
>
> (Rolfe cited in Gratton and Taylor, 1995, p. 226)

The success of these events depends, to a large extent, on the enthusiasm of the local communities and the capacity of the organizers, rather than on the natural and built heritage of the place (Getz, 1997).

Festivals that take place in rural areas can play an important role in economic terms (Calado, 2007). In fact, the development and organization of events are usually associated with tourism and the economic opportunities they can generate in a place. The economic impact results from the consumption of tourism products in a given geographical area; accommodation units, carriers, restaurants and shops are the main beneficiaries of the event. By attracting a large number of visitors, for a certain period of time, festivals favour an increase in average spending and can generate employment, albeit temporarily. In short, events are 'image builders' of the destination, they help to create a profile, position the destination in the market and provide it with a competitive advantage in terms of marketing (Getz, 1997).

The receiving communities have an important role in the development of events such as festivals and their success. It is also true, that associated mainly with festivals which are popular and where many people attend, fewer positive effects are generated, and instead negative impacts result from traffic congestion, rising prices of goods and services, the parking problem, vandalism or other factors that can make the relationship between guests and hosts difficult (Ferreira and Martins, 2007).

Dias (2003) identified eight categories of social impacts resulting from the type of tourism: (i) local resentment resulting from the shock of cultures; (ii) transformation of the work structure; (iii) the saturation of the infrastructure, which affects facilities and equipment; (iv) transformation of values and moral conduct; (v) modification of consumption patterns; (vi) health problems; (vii) ethnocentrism; and (viii) excess of standardization.

In fact, major festivals involve movements of people from different and distant locations, causing the establishment of social relationships between individuals who would not otherwise meet. The cultural shocks resulting from the difference are rooted in different values, lifestyles, religious beliefs, languages and different levels of prosperity; they tend to increase depending on the behaviour of visitors. Often, out of ignorance, or carelessness, visitors disrespect local customs and values. When this happens, violent reactions can be triggered. In fact, the populations of the destination place tend to develop stereotypes of the different communities that visit them, later adopting a standardized behaviour in the relationship with each one.

Because it is common to tourists and residents, the use of facilities and resources can have a serious impact on daily life, increasing the rejection and irritability of the local population in relation to visitors (Doxey, 1975; Butler, 1980). Destination saturation occurs, in general, when a rapid pace of development or market seasonality causes the coexistence, in the same space, of a higher number of tourists compared with the local population. When reaching the saturation point, problems can arise in terms of housing prices, employment (seasonality), transport (traffic congestion) and population (density of tourists). These factors, therefore, cause conflicts in the use of local resources, increase criticism regarding the deterioration of cultural heritage, and the traditional use of the territory changes (Butler, 1980).

However, tourism generates several benefits for those who practise it and for the receiving communities. Among the various social and cultural positive aspects of tourism, we underline: (i) increased information, acceptance and integration between different cultures promoted by social contact between visitors and locals; (ii) circulation of new ideas and values in the receiving community; (iii) a local culture can be renewed by the interest aroused in tourists; (iv) the presence of visitors leads to the creation and development of many useful services that make life easier for the local population; (v) there is an appreciation of the historical heritage; and (vi) intercultural exchange between different peoples increases dialogue, tolerance and understanding (Dias, 2003).

Knowing the perception of the inhabitants of the municipality of Idanha-a-Nova about the Boom Festival implies realizing that the type of destination and the level of tourism development in the territory are fundamental. In fact, the formation of perception results from the individual's social interactions with the environment. Reisinger and Turner (2003) define perception as the characterization that an individual constructs of a person, object or event involved in the surrounding context. In this sense, a positive perception of residents is generally observed when the economic and sociocultural benefits of tourism (e.g. improvements in income and tax revenues; increased employment; a greater number of recreational opportunities; increased levels of quality of life; cultural exchange; improvement of public and recreational services; enhancement of cultural heritage; revitalization of local traditions; and the increase in tourist infrastructure and equipment in destination regions (Mathieson and Wall, 1990; Sancho, 2001; OMT, 2003)) are perceived. On the other hand, economic and sociocultural costs (e.g. increase in the prices of goods and services; the increase in crime; the increase in road traffic; the increase in illegal games; the commodification of culture; the increase in prostitution; the spread of infectious diseases; and the increase in drug use (Mathieson and Wall, 1990; Sancho, 2001; OMT, 2003; Cooper *et al.*, 2007; Amuquandoh, 2009; Andereck and Nyaupane, 2011)) promote fewer positive perceptions.

Amuquandoh (2009) and Andereck *et al.* (2005) maintain that residents have also identified positive environmental effects of tourism, including: (i) conservation of the natural heritage; (ii) the preservation of species at risk; and (iii) the preservation of biodiversity. However, both studies as well as Archer and Cooper (2002) concluded that residents identified environmental costs such as air pollution, water pollution, noise, vandalism and wildlife destruction.

Brief Description of the Municipality of Idanha-a-Nova

The municipality of Idanha-a-Nova belongs to the district of Castelo Branco. It has a total area of 1412.7 km^2, bordering Spain to the south and east, to the north with the municipality of Penamacor, and to the west with the municipality of Castelo Branco. It is one of the larger municipalities in the region, and it has natural and scenic heritage of national interest. It is a member of the Naturtejo Geopark that is part of the United Nations Educational, Scientific and Cultural Organization's (UNESCO) worldwide network of natural parks.

Idanha-a-Nova was classified in 2015 as a City of Music, within the scope of UNESCO's Creative Cities Network; the first Portuguese locality to join the group. Since the beginning of the 1990s, Idanha-a-Nova has been developing a strategy that involves, among other areas, music.

According to data from Pordata (2019), Idanha-a-Nova has a population of 8100 individuals, distributed across 13 parishes and parish unions; 24.2% of the population resides in the village of Idanha-a-Nova. With an ageing population this space is characterized by isolation, rurality and low demography.

The county's ageing trends have been increasing at a faster rate than in Portugal: ageing rates are 427% in the municipality of Idanha-a-Nova and 161% in Portugal. These values obviously affect the ability to rejuvenate: the municipality of Idanha-a-Nova has a birth rate (5.7%) lower than that of Portugal (8.4%); and the mortality rate is higher (25.9%) than in the country (10.9%) (Pordata, 2019). However, between 2010 and 2019, there was an increase in the foreign population as a percentage of the resident population from 2.6% to 5.7%; in 2019, Idanha-a-Nova registered the same percentage as Portugal (5.7%) (Pordata, 2019).

In light of these features, how does the population of Idanha-a-Nova perceive summer music and dance events in terms of attracting visitors? What do they think about the Boom Festival?

The Boom Festival

The Boom Festival is a festival of visionary culture that takes place every 2 years, in the municipality of Idanha-a-Nova. It first took place in 1997, at Herdade da Azambuja, in the municipality of Setúbal; it has evolved, throughout its editions, from a musical event to a celebration of alternative culture.

In Idanha-a-Nova, the event took place in different locations: 'The festival had moved from its original forest location – where it stayed until the year 2000 – to a new lakeside spot in Herdade do Torrão in 2002, and where it continued to evolve until 2008' (Boom Festival Team, 2018, p. 33).

Boom is a multidisciplinary, transgenerational and intercultural festival; it crosses several artistic currents, such as painting, sculpture, land art, interactive installations, music, video art, plastic art gallery and graffiti (Boom Festival Team, 2008a).

That the event is multidisciplinary is supported by the festival poster which shows the conferences, workshops, gatherings and presentations of themes available during the event. There are approaches related to alternative science methodologies, exobiology, visionary culture, ancestral cultures, spirituality, Gnosticism, anthropology, psychology of conscience, paganism or emotional intelligence. It is transgenerational because it welcomes a diverse audience in terms of age; besides young people (teenagers and those in the 20s), the age group that most participates in this type of event, children and adults also go to the festival in large numbers (Boom Festival Team, 2008b). This diversity is achieved, according to the organizers, by the avant-garde artistic and cultural proposals by including hedonism and knowledge in the same event (Boom Festival Team, 2008b). The festival's interculturality comes from a vast worldwide network of ambassadors that attracts audiences from all continents. Contact between people from different cultural backgrounds is guaranteed; interaction in a context surrounded by different arts and culture, organizers maintain, can diminish differences or even favour the overthrow of interethnic stereotypes (Boom Festival Team, 2008b).

Boom is logo free; it is independent from the commercial system. In fact, unlike what happens at other summer festivals in Portugal (Super Bock Super Rock, Optimus Alive and Rock in Rio among others), the festival does not accept any sponsorship. The organizers intend to keep their space free of visual pollution from logos; they say that Boom's ethics are based on contact with nature in which the public is defended from any marketing strategy (Boom Festival Team, 2008b).

The Boom Festival's environmental concern has earned it several international awards. In this sense, and as an example, at the festival in 2016: (i) 90% of the electricity spent on camping was solar in origin; (ii) the cutlery used was 100% biodegradable; (iii) 550 trees were planted; (iv) 38,500 m^2 of gardens were created; (v) 80,400 kg of organic waste was made into compost; and (vi) 77% of the food consumed was vegetarian.

It was in 2010, already at Herdade da Granja, that the festival entered a new phase: Boomland was created which, according to the organizers, was 'a tailor-made Project created by people for people' (Boom Festival Team, 2018, p. 33). They state: 'It comprised a group of people with a common world view and set of social values' (Boom Festival Team, 2018, p. 33). Organizers add that, 'The physical natural environment enhanced the participation and cohesiveness of community members who lived either permanently or temporarily within Boomland or its surroundings' (Boom Festival Team, 2018, p. 33).

The ideal of the Boomland community was:

> Bringing life to the land .... Thousands of square meters of gardens and native trees have been planted since 2010. New lakes were designed. ... a number of proactive initiatives have enriched habitats and ecosystems, including the planting of over a thousand trees.
>
> (Boom Festival Team, 2018, p. 34)

In 2014, the Boom Karuna charity project was born. It was an initiative that 'set up local non-profit schemes which meet the festival's policy goals of supporting local suppliers and farmers while offering employment to the local population' (Boom Festival Team, 2018, p. 34).

In 2017, the Boom Festival organization acquired 150 ha in Herdade da Granja with the aim of developing other activities in Boomland.

The number of participants in the event has grown exponentially since it first took place up until 2018. In 1997, the participants in the festival were around 3500, while in 2018 there were 41,766. But it was from 2014 that the number of festival participants stabilized at around 40,000, due to issues related to the festival's sustainability and carrying capacity. The Boom was scheduled between 28 July and 4 August 2020, under the theme 'Anthropocene', but was not held due to the coronavirus disease 2019 (COVID-19) pandemic experienced in Portugal and around the world. Note, however, that ticket sales made online sold out in less than 1 hour (TSF Radio News, 2020); boomers from 60 countries purchased tickets. Similarly Boom Festival 2021 was postponed due to the pandemic and the festival will take place on 22–29 July 2022.

The Boom Festival generates a total economic impact in the country of €55.3 million, according to a study by Ernst & Young (Público, 2020). This value includes the sum of direct, indirect and induced impacts in the national territory. The news read, 'The event generates an added value of 29.4 million euros, mainly in the sectors of commerce and industry, having an indirect and induced impact of 30.1 million euros' (Público, 2020). In the tourism sector, the impact of the festival is €6.9 million. Non-resident boomers spend 7.4 days at the festival and 6.4 days at other destinations.

In addition, the number of people involved in the event, including festival goers, artists and other professionals, is 41,000, 'a value that represents 4.9 times the population of Idanha-a-Nova and is equivalent to about twice the number of tourists visiting the municipality for the year' (Público, 2020).

The Boom Festival and Deviant Behaviours

Those looking for summer festivals do so because of the culture of dance music that is usually associated with recreational contexts, collective lifestyles and forms of sociability of their own (Hitzler, 2002). Victor Silva (2005), in a study of dance culture in Portugal, divides this youth movement into three subcultures: (i) techno fans; (ii) house fans; and (iii) trance fans. While they share the essentials, each has its own set of practices and standards. The trance movement assumes a set of values and ideological principles that, intentionally, are opposed to the current model of society; in particular, it has its night parties outdoors in non-urbanized areas, in the countryside, in forests, in short, in areas far from inhabited places (Calado, 2007).

The trance party can be seen as a collective celebration of a subculture, which celebrates a ritual with mystical elements and in communion with nature, in the form of dance and playful 'trance'. Contributing to this collective feeling is the fact that the participants are in pairs, together with other individuals who share the same values, away from public view. The trance party functions as a 'counter-daily life', an event delimited in time and space. The notion of 'escape' from the day to day, from the real world, permeates the entire discourse of the trancers and culminates in the night party. This escape is, at all times, instigated in the most varied ways, in the form of an exacerbated hedonism (Calado, 2007).

As some authors have pointed out (Horta, 1997; Carvalho, 2003), the trance subculture is based on a retelling of the psychedelic era from the mid-1960s to mid-1970s, updating the ideas of hippy and other pacifist utopias. At the same time, it fuses mystical, esoteric, environmentalist, New Age and other elements.

The psychedelic movement is marked by the consumption of products that cause various hallucinations, distort the senses, and stage the release of serotonin or dopamine. The escape from everyday life, the search for artificial sensory havens and the intentional alteration of the senses are, in these cases, ends in themselves (Calado, 2007).

How did the residents of Idanha-a-Nova perceive this clash between lifestyles? A rural community is confronted with a group of visitors who reveal irreverent, heterodox, transgressive and liberating practices. What is the reaction? Thus, we analysed the results of the surveys carried out in 2009 and 2021, on the differences and similarities in the perceptions of the people from Idanha-a-Nova regarding the Boom Festival.

Method, Instrument and Samples

The study was carried out using two samples from residents in the municipality of Idanha-a-Nova:

1. The first sample consisted of 389 residents who were surveyed between December 2008 and April 2009. Respondents answered a questionnaire in the form of an interview, on the street, after being informed about the objective of the study. Freedom of participation and confidentiality of the requested information were guaranteed.
2. The second sample consisted of 134 residents who were surveyed between September 2020 and January 2021. Considering all the constraints caused by the COVID-19 pandemic, respondents answered the online questionnaire after disclosing the link to access residents. Information was provided regarding the purpose of the study, and freedom of participation and the confidentiality of the requested information were guaranteed.

The data collection instrument used was specially developed for this purpose, consisting of 12 questions, in addition to questions of a sociodemographic nature (gender, age, composition of the household, activity and literacy of respondents). With these questions, we wanted to know the respondents' perceptions of the Boom Festival.

The statistical analysis of the data obtained was performed with version 27.0 of the SPSS software.

Table 10.1 contains the information regarding the characterization of the samples under study. Both samples are mainly made up of female respondents, aged over 50, married and with children. In 2009, most respondents did not have children under the age of 30 living with them, while in 2021 this trend is reversed, highlighting the consequences of the pandemic and economic crisis that Portugal is going through. Both samples are mostly made up of residents of the village of Idanha-a-Nova with employment. Regarding educational qualifications, most of the respondents in 2009 had completed further education while in 2021 most respondents had received higher education.

Results

Table 10.2 shows the results relating to the knowledge and participation of respondents in the Boom Festival. Between 2009 and 2021 there was a decrease in the number of respondents who had never heard of the Boom Festival, from 17.2% to 4.5%. The percentage of respondents who participated in the Boom Festival increased from 42.4% to 60.2%, with most of the respondents participating as visitors (58.2% in 2009 and 75.3% in 2021) showing an increase in interest in this event. The main source of publicity for the Boom Festival was, in both times, family and friends. The perception that this is an international event gained more expression in 2021.

Figure 10.1 shows the frequency distributions of the items related to the question 'What can we find/watch at the Boom Festival?' in 2009 and in 2021. For all aspects, there was a more correct perception in 2021 of what can be found at the Boom Festival.

Respondents were asked to rate the Boom Festival on a scale of 1 to 10 (Table 10.3). The data show a more positive perception in 2021, with a greater homogeneity of responses by the respondents (lower standard deviation).

When challenged to mention a word that synthesizes (i.e. sums up) the image, they have of the Boom Festival (Fig. 10.2), most respondents referred to 'music' both in 2009 (31.86%) and in 2021 (25.0%). In 2009, 'social' (13.56%) and 'drugs' (13.56%) also assumed some expression, while in 2021 the ideas 'funny' (17.19%) and 'different/alternative' (17.19%) stood out.

Respondents were also asked to indicate a positive aspect (Fig. 10.3) and a negative aspect (Fig. 10.4) for the Boom Festival. Both in 2009 and in 2021, the main positive aspect was the economic impact that this event has on the local economy, as well as the fact that it was an important means of disclosing Idanha-a-Nova.

Table 10.1. Sociodemographic information about the two samples of Idanha-a-Nova residents (2009 and 2021) studied. Compiled by the authors from survey data.[a]

	Total		2009		2021	
	N	%	N	%	N	%
Gender						
Female	288	55.1	205	52.7	83	61.9
Male	235	44.9	184	47.3	51	38.1
Age (years)						
Less than 18	16	3.1	15	3.9	1	0.7
18–25	92	17.6	78	20.1	14	10.4
26–35	91	17.4	76	19.5	15	11.2
36–49	125	23.9	82	21.1	43	32.1
50 or more	199	38.0	138	35.5	61	45.5
Marital status						
Single	188	35.9	141	36.2	47	35.1
Married	246	47.0	188	48.3	58	43.3
Cohabiting	14	2.7	3	0.8	11	8.2
Divorced	37	7.1	24	6.2	13	9.7
Widowed	38	7.3	33	8.5	5	3.7
Have children?						
No	222	42.4	163	41.9	59	44.0
Yes	301	57.6	226	58.1	75	56.0
Children under 30 living with you						
No	217	61.6	117	53.7	25	42.4
Yes	135	38.4	101	46.3	34	57.6
Parish						
Idanha-a-Nova	323	61.8	223	57.3	100	74.6
Other	200	38.2	166	42.7	34	25.4
Activity						
With employment	300	57.4	208	53.5	92	68.7
Looking for first job	8	1.5	8	2.1	–	–
Unemployed	30	5.7	25	6.4	5	3.7
Retired/pensioner	106	20.3	81	20.8	25	18.7
Student	79	15.1	67	17.2	12	9.0
Academic qualifications						
No studies	37	7.1	33	8.5	4	3.0
Primary school	81	15.5	76	19.5	5	3.7
Secondary school	91	17.4	72	18.5	19	14.2
Further education	172	32.9	135	34.7	37	27.6
Higher education	142	27.2	73	18.8	69	51.5

[a]Total sample sizes were 389 residents (2009) and 134 residents (2021).

Table 10.2. Results regarding respondents' knowledge and participation in the Boom Festival for the total sample, 2009 and 2021.

	Total		2009		2021	
Question	N	%	N	%	N	%
Have you heard of the Boom Festival?						
Yes	450	86	322	82.8	128	95.5
No	73	14	67	17.2	6	4.5
Have you ever attended the Boom Festival?						
Yes	213	47.4	136	42.4	77	60.2
No	236	52.6	185	57.6	51	39.8
What was your participation due to?						
Visitor	140	64.2	82	58.2	58	75.3
Professional reasons	73	33.5	54	38.3	19	24.7
Other	5	2.3	5	3.5	–	–
How did you hear about the Boom Festival?						
Family/friends	195	43.7	129	40.6	66	51.6
For the movement in the village	88	19.7	62	19.5	26	20.3
Internet	31	7	18	5.7	13	10.2
Social communication	96	21.5	74	23.3	22	17.2
Other	36	8.1	35	11	1	0.8
In your opinion, this is a festival/event						
Local	11	2.5	9	2.8	2	1.6
Regional	9	2	9	2.8	–	–
National	18	4	14	4.4	4	3.1
International	407	91.5	285	89.9	122	95.3

About the negative aspects mentioned, the respondents' perceptions changed significantly during the period under analysis. Excessive consumption was the main negative aspect in 2009 (39.42%) followed by traffic (27.24%). In 2021 the respondents mentioned as the main negative aspect the price increase verified in the local shops during the festival days (43.65%) followed by excessive consumption (33.33%).

Respondents were presented with a set of statements regarding the Boom Festival asking them to express their agreement or disagreement. Figure 10.5 compares the average values obtained for each of the items under analysis in 2009 and in 2021.

The Mann-Whitney test allowed the identification of statistically significant differences for the items:

- 'It brings many visitors to the region' ($p < 0.001$);
- 'It is positive for the village's economy' ($p < 0.001$);
- 'It brings more noise to the village' ($p < 0.001$);
- 'Causes traffic congestion' ($p < 0.001$);
- 'It causes more pollution in the village' ($p < 0.001$);
- 'It makes the village more insecure' ($p < 0.001$); and
- 'Changes the identity of the inhabitants of Idanha' ($p = 0.013$).

For all items, lower average scores were recorded in 2021, reflecting a devaluation of positive aspects such as the number of visitors arriving in Idanha-a-Nova and the respective impact on the local economy. However, there is also a significant

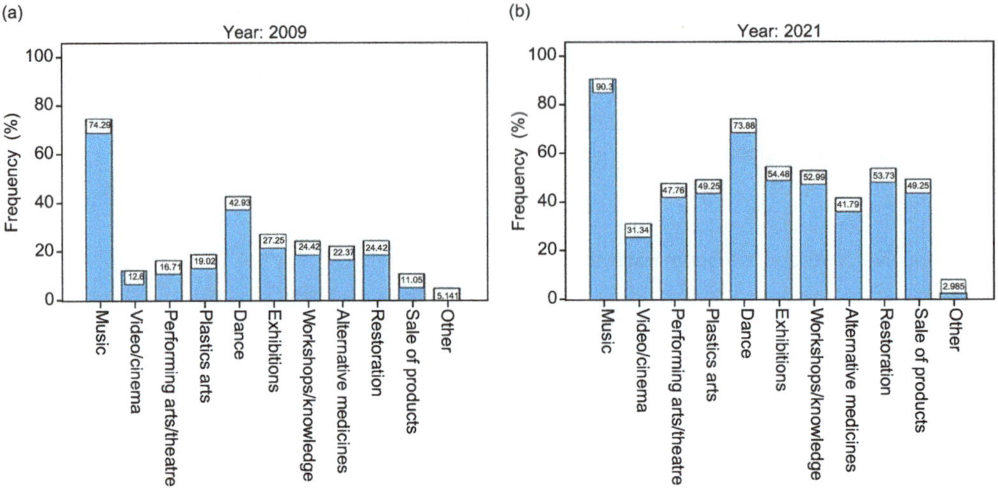

Fig. 10.1. Frequency distributions of items related to the question 'What can we find/watch at the Boom Festival?' in (a) 2009 and (b) 2021.

Table 10.3. Basic descriptive statistics for the item 'Overall, score the event from 1 to 10 (where 1 is the minimum score and 10 the maximum score)'.

Global score from 1 to 10	N	Minimum	Maximum	Mean	Standard deviation
Total	446	1	10	7.44	2.262
2009	318	1	10	7.21	2.289
2021	128	1	10	8.03	2.088

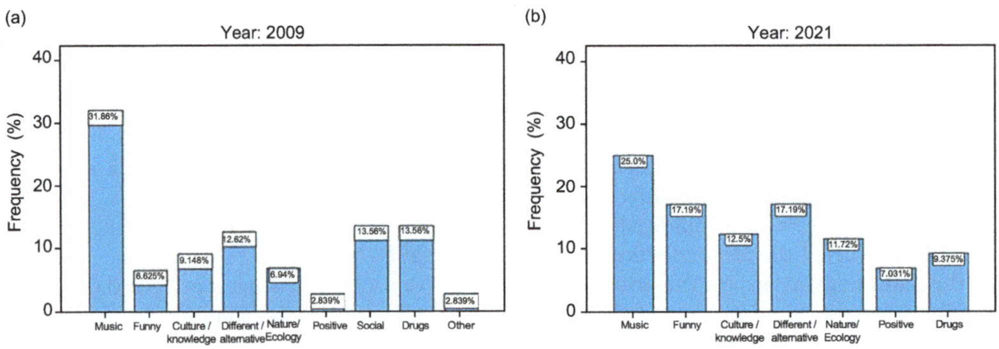

Fig. 10.2. Frequency distributions of items related to the question 'Refer to a word that synthesizes the image of the Boom Festival' in (a) 2009 and (b) 2021.

devaluation on the part of respondents of aspects related to the negative effects on traffic, pollution, safety and the identity of local residents.

When asked whether they consider the Boom Festival to be welcome in Idanha-a-Nova and whether they would like the Boom Festival to remain in Idanha-a-Nova, the overwhelming majority of respondents answered in the affirmative (Table 10.4) both in 2009 as in 2021. Still, it is worth noting the fact that these percentages were even more expressive in 2021, exceeding 90%.

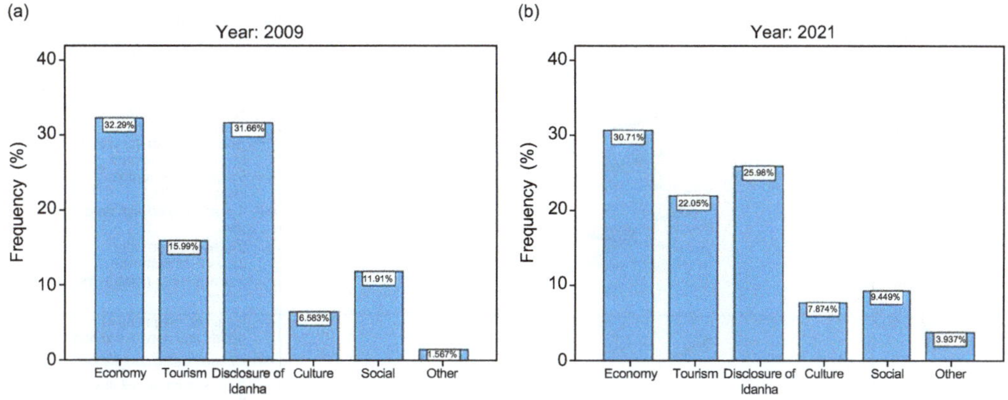

Fig. 10.3. Frequency distributions of items related to the question 'Indicate a positive aspect of the Boom Festival' in (a) 2009 and (b) 2021.

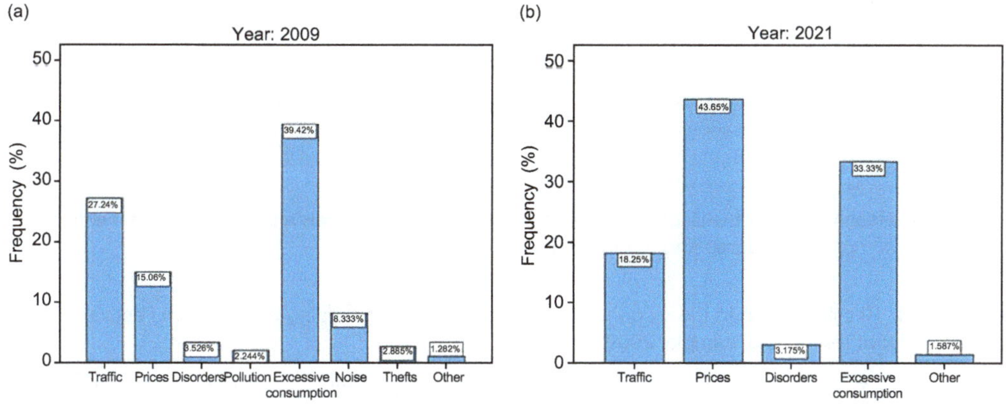

Fig. 10.4. Frequency distributions of items related to the question 'Indicate a negative aspect of the Boom Festival' in (a) 2009 and (b) 2021.

Also, about the image of the Boom Festival, respondents considered that there has been an improvement over time, with this perception being more pronounced in 2021 (Table 10.5). In 2021 there was not only a higher average score, but also a lower standard deviation, reflecting more agreement on the part of respondents.

Final Considerations

Tourism has been an important engine of economic development in many communities with rural characteristics. For the success and sustainable development of the tourist activity, the involvement of all intervening subjects and, in particular, the local community is of great importance. In this sense, it is important to assess the residents' perceptions regarding the development of these activities.

The study carried out here concluded that the residents of Idanha-a-Nova perceive the Boom Festival in a very positive way, highlighting the positive impacts more than the negative ones. The positive evolution observed in the perception of residents between 2009 and 2021 stands out.

Between 2009 and 2021 there was an increase in the residents' knowledge about the Boom Festival. The number of residents who

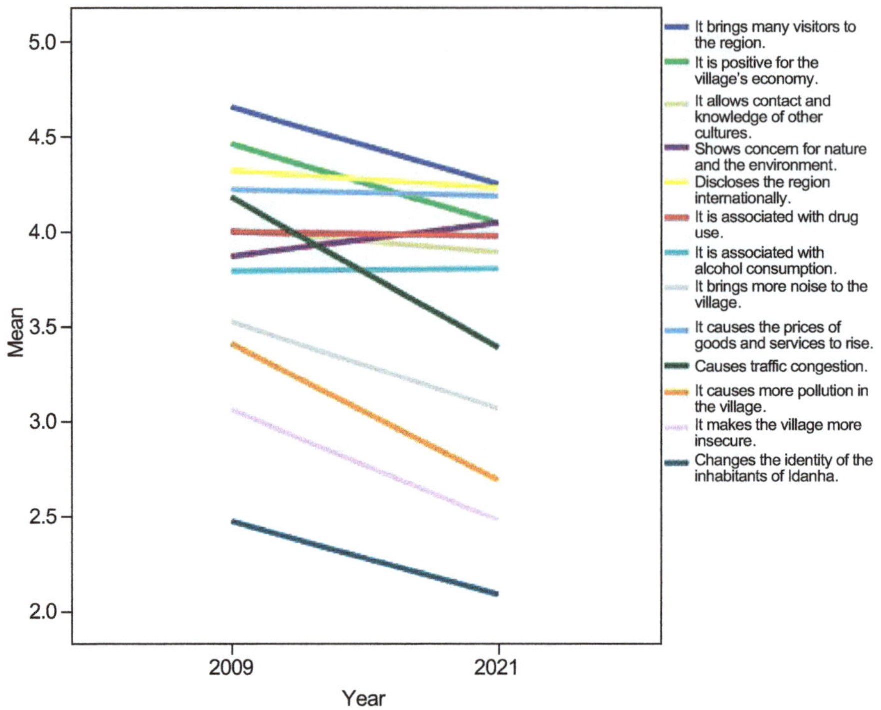

Fig. 10.5. Mean values for the degree of agreement for a set of statements about the Boom Festival where 1 is completely disagree and 5 totally agree in 2009 and 2021.

Table 10.4. Results related to the respondents' opinions about the holding of the Boom Festival in Idanha-a-Nova for the total sample, 2009 and 2021.

	Total		2009		2021	
Question	N	%	N	%	N	%
In your opinion, is the festival welcome in Idanha-a-Nova?						
Yes	383	89.7	266	89.0	117	91.4
No	44	10.3	33	11.0	11	8.6
Would you like the Boom Festival to remain in Idanha-a-Nova?						
Yes	363	89.6	245	88.4	118	92.2
No	42	10.4	32	11.6	10	7.8

Table 10.5. Basic descriptive statistics for the item 'To what extent, has the image of the Boom Festival been improving, year by year, for people living in the municipality of Idanha-a-Nova?' (where 1 means 'no improvement' and 10 means a 'a lot').

Global score from 1 to 10	N	Minimum	Maximum	Mean	Standard deviation
Total	440	1	10	6.42	2.836
2009	312	1	10	5.99	2.846
2021	128	1	10	7.47	2.531

have already participated in this event also increased, in particular an increase in those who participated as visitors.

Regarding the positive impacts perceived by residents, the main positive aspect, both in 2009 and in 2021, stands out as being the economic impact that this event has on the local economy as well as the fact that it is an important means of publicizing Idanha-a-Nova. With regard to the negative aspects mentioned, the respondents' perceptions changed significantly during the period under analysis. Excessive consumption is the main negative aspect in 2009, while in 2021 the respondents mentioned as the main negative aspect the price increase seen in local shops during the festival days.

In general, residents consider that the Boom Festival is welcome in Idanha-a-Nova, and they want this event to continue to take place in their municipality and consider that the image of the Boom Festival has been improving significantly over the years.

References

Abreu, P. (2004) Músicas em movimento. Dos contextos, tempos e geografias da performance musical em Portugal. *Revista Crítica de Ciências Sociais* 70, 159–181.

Amuquandoh, F.E. (2009) Residents' perceptions of the environmental impacts of tourism in the Lake Bosomtwe Basin, Ghana. *Journal of Sustainable Tourism* 18(2), 223–238.

Andereck, K.L. and Nyaupane, G.P. (2011) Exploring the nature of tourism and quality of life perceptions among residents. *Journal of Travel Research* 50(3), 248–260. DOI: 10.1177/0047287510362918.

Andereck, K.L., Valentine, K.M., Knopf, R.C. and Vogt, C.A. (2005) Residents' perceptions of tourism related community impacts. *Annals of Tourism Research* 32(4), 1056–1076. DOI: 10.1016/j.annals.2005.03.001.

Archer, B. and Cooper, C. (2002) Os impactos negativos e positivos do turismo. In: Theobald, W. (ed.) *Turismo Global*. Senac, São Paulo, Brazil, pp. 85–102.

Boom Festival Team (2008a) Boom Festival 08 – Press Release. Available at: http://www.boomfestival.org/boom2008/press/PressRelease_print.pdf (accessed 10 February 2009).

Boom Festival Team (2008b) *Boom Book*. Good Mood Lda, Idanha-a-Nova, Portugal.

Boom Festival Team (2018) *Boom Festival 20 Years: a Visual and Oral History*. Good Mood Lda, Idanha-a-Nova, Portugal.

Butler, R.W. (1980) The concept of a tourist area cycle of evolution: implications for management of resources. *Canadian Geographer* 24(1), 5–12. DOI: 10.1111/j.1541-0064.1980.tb00970.x.

Calado, V.G. (2007) Trance psicadélico, drogas sintéticas e paraísos artificiais. Representações: uma análise a partir do ciberespaço. *Toxicodependências* 13(1), 21–28.

Carvalho, M.C. (2003) Expressões psicadélicas juvenis. In: Cordeiro, G.I., Baptista, L.V. and Firmino da Costa, A. (eds) *Etnografias Urbanas*. Celta, Oeiras, Portugal, pp. 167–178.

Cooper, C., Fletcher, J., Fyall, A., Gilbert, D. and Wanhill, S. (2007) *Turismo: Princípios e Práticas*, 3rd edn. Bookman, Porto Alegre, Brazil.

Dias, R. (2003) *Sociologia do Turismo*. Editora Atlas, São Paulo, Brazil.

Doxey, G.V. (1975) When enough's enough: the natives are restless in Old Niagara. *Heritage Canada* 2, 26–27.

Ferreira, A.M. and Martins, A.I. (2007) *O Evento FCNC 2005 e o Turismo*. Universidade do Algarve, Faro, Portugal.

Getz, D. (1997) *Event Management and Event Tourism*. Cognizant Communication Corporation, New York.

Gratton, C. and Taylor, P.D. (1995) Impacts of festival events: a case study of Edinburgh. In: Ashworth, G.J. and Dietvorst, A.G.J. (eds) *Tourism and Special Transformations for Policy and Planning*. CAB International, Wallingford, UK, pp. 225–238.

Hitzler, R. (2002) Pill kick: the pursuit of "ecstasy" at techno-events. *Journal of Drug Issues* 32(2), 459–465. DOI: 10.1177/002204260203200208.

Horta, H. (1997) As dinâmicas sócio-culturais das re-uniões associadas à música trance. Tese de Licenciatura, Universidade Autónoma de Lisboa, Lisboa, Portugal.

Janiskee, B. (1980) South Carolina's harvest festivals: rural delights for day tripping urbanites. *Journal of Cultural Geography* 1(fall/winter), 96–104. DOI: 10.1080/08873638009478655.

Mathieson, A. and Wall, G. (1990) *Tourism: Economic, Physical and Social Impacts*. Longman Scientific & Technical, London.

OMT (Organização Mundial de Turismo) (2003) *Turismo Internacional: Uma Perspetiva Global*, 2nd edn. Bookman, Porto Alegre, Brazil.

Pordata (2019) O seu município em números. Available at: https://www.pordata.pt/Municipios/Quadro+Resumo/Idanha+a+Nova-252286 (accessed 20 January 2021).

Público (2020) Boom Festival tem um impacto de 55,3 milhões. Available at: https://www.publico.pt/2020/02/06/p3/noticia/boom-festival-impacto-economico-553-milhoes-euros-1903093 (accessed 20 January 2021).

Reisinger, Y. and Turner, L.W. (2003) *Cross-Cultural Behavior in Tourism*. Elsevier Butterworth-Heinemann, Oxford, UK.

Ribeiro, J.C., Vareiro, L.C., Fabeiro, C.P. and de Blas, X.P. (2005) Importância da celebração de Eventos Culturais para o Turismo do Minho-Lima: um estudo de caso. Available at: http://www3.eeg.uminho.pt/economia/nipe/docs/Publicacoes_Outras_Revistas/Com%20Arbitragem/2006/Ribeiro_Vareiro_Fabeiro_Blas_2006_RPER.pdf (accessed 20 February 2009).

Sancho, A. (2001) *Introdução ao Turismo: Organização Mundial do Turismo*. Roca, São Paulo, Brazil.

Silva, V. (2005) Techno, house e trance. uma incursão pelas culturas da 'dance music'. *Toxicodependências* 11(2), 11–19.

TSF Radio News (2020) Boom Festival com impacto económico de 56 milhões de euros em Idanha-a-Nova. Available at: https://www.tsf.pt/portugal/cultura/boom-festival-com-impacto-economico-de-56-milhoes-de-euros-em-idanha-a-nova-11783245.html (accessed 23 January 2021).

11 Community-initiated Event Tourism in the Global South: the Case of South Africa and the Kingdom of Lesotho

Lwazi Apleni[1], Jankie Kgalabi Phale[2] and Unathi Sonwabile Henama[2]*

[1]University of Zululand, KwaDlangezwa, South Africa; [2]Tshwane University of Technology, Pretoria, South Africa

Introduction

The rapid growth of South Africa can be associated with the end of apartheid and the end of segregation policies in 1994, which ushered in the first democratic elections. This began the growth phase of tourism which had remained primarily anti-developmental during the era of apartheid, as it was not a fashionable destination to visit. Ensuring shared growth in the tourism industry has been an area of focus to ensure that tourism is impactful in addressing the structural challenges facing the South African economy. Poverty, unemployment and inequality remain stubborn structural challenges, which tourism has great potential to resolve. The tourism industry has grown in leaps and bounds, its growth rates far exceeding the gross domestic product (GDP) growth of the economy. In the majority of instances, tourism had achieved double-digit growth rates, for a country that had struggled since 1994 to breach GDP growth rates of 5% per year. Tourism has gained prominence as it can be an important catalyst for post-production economic activity, as it is a service and an experience located within the consumption economy. This means that outlying areas, outside of the urban centres associated with urban tourism, can also benefit from tourism if they package themselves for tourism consumption. This is the experience of countries located within the Global South, represented by developing countries that seek to increase their global share of international tourism trade.

The study described in this chapter used a review of the literature to investigate events that have been promoted in rural areas and that have benefitted the geographic spread of tourism. These community-initiated events have grown organically, attracting more visitors and aspects of commercialization, and have grown to become hallmark events in their localities. Semonkong, in landlocked Lesotho, is a festival, premised on music and entertainment, where camping is the theme of the festival and the primary accommodation for the vast majority of festival goers. The town of Berlin, outside East London, has a community-initiated traditional horse racing festival that has gained prominence and has become a major festival that has developed and achieved partial commercialization. The community of Hammanskraal outside Pretoria had been a recruitment site for many Africans who participated in World War I in support of Britain. Upon a return to South Africa, the former soldiers

*Corresponding author: HenamaUS@tut.ac.za

DOI: 10.1079/9781789248685.0011

would undertake their drills and the community would participate and this developed organically to become a major festival during the month of December. The listed festivals represent small-scale festivals that are locally initiated and have a direct impact on the local citizens and business community. They are generally small scale in nature and have not attained full-scale commercialization and potentially may not meet the gaze of mass tourism.

Many developing countries have jumped on the tourism bandwagon, trying to secure the positive economic impacts associated with tourism. Tourism to Africa is disproportionally important considering its huge dependence on natural resource exports and low levels of beneficiation. Tourism is a major attraction of foreign exchange for many developing countries, as tourism is a major export industry consumed at the destination area. Africa is one of the fastest-growing regions in terms of tourism, even though Africa represents a small share of the global tourism receipts. 'Tourism is an export industry, where the tourism product offering can be consumed at the destination area which means that the bulk of the value adding happens at the destination area' (Henama, 2017, p. 72). The growth of tourism in recent years has been associated with a megatrend, namely, the growth of millennials (i.e. people reaching young adulthood in the early 21st century). Millennials are one of the major consumer groups that have impacted on the demand for tourism, as they seek more authentic experiences. Apleni and Henama (2020) elucidate that the purpose of events is typically economic, that is to generate or build the local economy through attracting inward investment, new businesses or increased tourist visitation and expenditure, though of course sociocultural benefits may also accrue through, for example, improved infrastructure and amenities. Internally, the purpose of events is often primarily sociocultural, to celebrate or strengthen local culture and, as a consequence, to enhance a sense of identity among local communities (Apleni and Henama, 2020). Events that are locally curated by the host community are more sustainable in nature as they are locally owned and would be hosted yearly instead of mega-events that hardly produce any long-term benefits especially for developing countries. Strydom *et al.* (2006) noted that the growth in tourism in South Africa during the past 10 years resulted in the development of various types of tourism such as ecotourism, cultural tourism, adventure tourism, business tourism, sports tourism and event tourism.

Theoretical Framework: Event Tourism

'Events are an important motivator of tourism and figure prominently in the development and marketing plans of most destinations' (Getz, 2008, p. 4). Events attract tourists to a destination area as they pull tourists like attractions to a destination they would otherwise not visit. Events can assist in the place marketing of a destination by creating positive images of a destination. Events can have a positive economic impact on a host community. Locals can develop community pride because of the event and this is one of the social benefits of event tourism. Events could be planned to take place during the shoulder season of a destination, thereby attracting tourists to address seasonality. According to Etiosa (2012), event tourism is a systematic planning, development and marketing of festivals and special events as tourist attractions, image makers, catalysts for infrastructure and economic growth and animators of build attraction. The sheer number of events has led to the development of the event tourism industry. South Africa has hosted major events such as the 2010 Fédération Internationale de Football Association (FIFA) World Cup but such mega-events hardly bring any positive economic impacts to developing countries. As Giampiccoli *et al.* (2013) observed:

> The findings from the various reports and figures from South Africa and FIFA show that the 2010 FIFA World Cup was a short term success as it brought a brief tourism boom to South Africa between June and July 2010. Contrary to the hope that the 2010 FIFA World Cup could have benefitted South Africa more widely, related tourism benefit has intensified only in Johannesburg, Cape Town, Durban and other major game parks and beaches near these centres. Government agencies have done little to promote rural tourism or to spread the tourism benefits more widely.
>
> (Giampiccoli *et al.*, 2013, p. 16)

Mirele (2015) noted that the expected benefits from the 2010 FIFA World Cup were not met. Smaller scale events are more sustainable for destinations when they are not one-off events like mega-events. This is because they can use existing infrastructure and possibly create an incentive for the state to create supportive infrastructure that will primarily benefit the local community in areas that are not regarded as tourism zones, located in peripheral areas. When events are hosted in rural and peripheral areas, they can become important economic boosters for these areas of low-density population, whose primary economic functioning can be linked to agriculture.

There are various types of events, ranging from special events, hallmark events, mega-events and local events. According to Etiosa (2012), events have the ability to promote the destination, facility or organization and attract tourists. This means that the hallmark event and destination become inseparable and the hallmark event would be closely associated with the destination and its image. 'Hallmark events are identified with the spirit and soul of a host community [so] that they become synonymous with the name of the place, and gain widespread recognition and awareness' (Oklobdzija, 2015, p. 86).

Details of some community-initiated events, along with observations of how these have benefitted the spread of tourism in rural areas of South Africa and the Kingdom of Lesotho, are provided below.

Berlin Traditional Horse Racing, Eastern Cape Province, Republic of South Africa

Berlin is a town located in the Eastern Cape Province of South Africa, which hosts the Berlin traditional horse racing event, which is an example of a successful community-initiated event. ShowMe (2019) noted that just 40 km outside East London is the town of Berlin, where the Berlin traditional horse racing event that has grown from 5000 visitors in its first year, to a very respectable 30,000 just 5 years later, has become a major tourist attraction. It is the only traditional horse racing event of its kind in the Eastern Cape. There is an existing tradition of horse riding among the Xhosa people that dominate the Eastern Cape Province in South Africa. While traditional horse racing is popular in the Eastern Cape's rural villages, where races are held regularly on weekends, the Berlin November festival is the biggest and richest of these meetings in the province so far. This created the means-to-an-end to develop the traditional horse riding event, in a destination that has horse races every weekend. The horses, which range from thoroughbreds to traditional Xhosa breeds and cross-breeds, are inspected by veterinarians from the government and the Society for the Prevention of Cruelty to Animals to ascertain if they have a clean bill of health to take part in the day's events (Ledwaba, 2017). Ledwaba (2017) elucidates that at its inception in 2013, the Berlin November drew a crowd of about 5000; however, in 2017 an estimated 30,000 fans turned up. The economic contribution of the Berlin November event is projected to be R23 million to the Eastern Cape economy, as the event is increasingly drawing people from outside the province.

As the event has grown yearly in numbers, it has attracted business opportunities where entrepreneurs now set up marquees and sell hospitality packages for the discerning members of society. This has led to an increase in tourists to the area. These developments mean that the event is changing in its character to attract a greater and a more diverse client base by also having hospitality packages sold. The event is attracting a non-traditional audience as the event is seen as an opportunity for corporates to entertain their clients, reward their staff, or as a year-end function opportunity with the trappings of live bands, the attendance of celebrities, a fashion parade, a business village with food courts and children's facilities. The event has developed and attracted the attention of politicians, which has led to financial and logistical support from the Eastern Cape Provincial Government. Clark (2016) noted that the Eastern Cape Department of Sports and Recreation has been part of the local government assistance for the Berlin November by installing toilet facilities, setting up a traditional food court and enlisting various music entertainers at the event.

The Eastern Cape Gambling and Betting Board, which is empowered through legislation over gambling and betting, got involved in regulating what was previously unregulated betting on traditional horses. The Berlin November is a community-initiated event that is located in the rural hinterlands of the Eastern Cape, and is premised on horse racing, which is practised as a sport on weekends across many localities in the Eastern Cape. The Berlin November is therefore the highlight and can be regarded as the 'World Cup' of traditional horse riding in the Eastern Cape. According to Umhlobo Wenene (2017, p. 1), 'the anchor of the Berlin November is the traditional horse racing handicap with a total prize money of R600,000, featuring different races'. The initial prize money for the winning horse was R50,000 and the involvement of state organs has led to the prize money increasing to R600,000. Interest in horse riding has increased far and wide in the Eastern Cape due to regional horse races and the Berlin November.

The event is successful in promoting the geographic spread of tourists into rural areas, which directly benefits the host community. Promoting the geographic spread of tourists into rural areas is one of the major objectives of tourism planning in South Africa as stated in the National Tourism Sector Strategy (NTSS). The event is located in the rural hinterland of the Eastern Cape, and is an example of a rural event. The event will continue to grow, as its appeal extends to other provinces and there is a hope that it might be the Eastern Cape's version of the Durban July, but located in a rural setting with a comparable economic impact.

Maletsunyane Braai Festival, Semonkong, Kingdom of Lesotho

The Kingdom of Lesotho is completely surrounded by South Africa, and is therefore a landlocked country with Maseru as its capital city. It was established by King Moshoeshoe I, and was annexed by the British, and is known for its rich diamond deposits. Lesotho is mountainous, and has the highest peak in Africa. According to Cycloscope (2019), Lesotho is the only state in the world completely above 1000 m; to be precise 80% of it is located above 1800 m. Cycloscope (2019) noted that the town of Semonkong, about 5 km from Maletsunyane Falls, is 2275 m above sea level. Cycloscope describes the waterfall as follows:

> With its 192 meters (630 feet) it is one of the most impressive waterfalls in Africa and among the highest single-drop falls, it plunges into the beautiful gorge below offering a breathtaking view. If you are brave enough, you can tackle the abseiling. At 204 meters, this is the longest rope descent in the world, entered into the *Guinness World Records* in October 2005.
>
> (Cycloscope, 2019, p. 1)

According to Somicimbi (2017), Semonkong – 'Place of Smoke' – was established in the 1880s as a refuge for Basotho displaced by the Gun War. Lonely Planet notes:

> Semonkong (Place of Smoke), a one-pony town in the rugged Thaba Putsoa range, gets its name from the nearby Maletsunyane Falls (240 m), which are at their loudest in summer. The town is the starting point for many fine hiking, and pony-trekking trails, including the two-day ride via the peaks of Thaba Putsoa to Ketane Falls (122 m).
>
> (Lonely Planet, 2015, p. 485)

The Maletsunyane waterfall provides the perfect environment to host a braai festival, which means a barbeque festival. Maletsunyane Braai Festival is an annual event happening during the last weekend of November and is the biggest braai festival in Lesotho. It has become a bucket-list event to attend, especially among adventure tourists because the Maletsunyane waterfall is one of the most impressive waterfalls in sub-Saharan Africa. The rural nature of the location means that there is scant commercial tourism accommodation, as there is only one hotel, Semonkong Lodge, which is always fully booked in advance for the duration of the braai festival. This has resulted in the framework of the braai festival being developed into a camping festival, where participants have two entry tickets, one for the event and one for the camping site. With camping being the basic option for the festival, located near the edge of the waterfall, it means that it is a responsible tourism event, as the camping tents and equipment are removed after the event and the site goes back to its natural state, uninhabited by humans. The character of the festival has allowed direct and indirect benefits for the local community from the expenditure of tourists and patrons that come to the braai festival. The World Economic Forum (2018) noted that Lesotho attracts around 320,000 tourists/year, and tourism contributes 5.8% of GDP. This is a far cry from the around 10 million tourists that neighbouring South Africa attracts, and the Maletsunyane Braai Festival is an opportunity to grow the number of South Africans visiting Lesotho.

Ditrupa (Hammanskraal), Gauteng Province, Republic Of South Africa

Hammanskraal is a peri-rural area located outside Pretoria, the administrative capital city of South Africa. Hammanskraal is a dormitory town, which provides labour for the administrative capital city, and basically has scant economic activity. In the last week of December each year, the town comes to life as the festive season experiences inward migration as people travel back to the rural hinterlands, returning to their ancestral lands and second homes. 'The Bakgatla ba ga Makapane says the annual "Diturupa" celebrations play an important role in social cohesion. The Makapanstad community and surrounding villages come together in song, drums and dance to celebrate the end of the year' (SABC, 2019b). SABC News (2018) noted that in the year 2018 an estimated 4000 people attended to watch around 30 groups perform at an open field, not far from the Makapan royal homestead. The Ditrupa Festival is a community-initiated event to commemorate the participation of Black South Africans as soldiers during World War I under the Union of South Africa. The festival was initiated by soldiers who returned home after World War I, as they would perform some army-related activities to show off their skills and this became a means to entertain the community, and it grew to become a festival according to SABC News (2019a, p. 1). According to SABC News (2019b), the event was started 101 years ago by World War I combatants during their holiday breaks from foreign missions as they visited their village homes. As they were languishing at home in boredom, some among them decided that they should start performing the military drills that they had learnt during the war.

The former soldiers formed troops according to whatever battalion they had been part of in the war and then starting drilling, to the amusement of their families and villagers at large. The tradition literally grew from those small beginnings to the extent where villagers at large took ownership of whatever the returning soldiers had taught them as drill exercises. Young men would then take over from their fathers, including the military gear that was used; this tradition must be applauded for not being patriarchal as women have always been a part of it – you always see them clad in their Scotch skirts (kilts) as mainline participants alongside men in their drills. This tradition has gone a long way in bringing together villagers from all walks of life and has made families find solace among each other, as in the real military sense of brothers in arms. On the 26 December, Makapanstad village hosts the biggest Ditrupa Festival with 40+ groups participating in the celebration and scores of supporters and fans driving from as far as Rustenburg, Vaal and many other far-flung areas to experience this day with the locals. The official start of the event is 10h00 but people arrive as early as 08h00 to reserve the best spot at the local ground where the festival happens. As it is, there are no official statistics of attendants of the festival but one will be forgiven to estimate that over 7000 people attend this festival. The North West provincial Department of Arts, Sports and Culture lists this event as one of their top events with the Member of the Executive Council attending in person in 2018. The organizers also give the required security priority to the event with a mobile police station on site, and emergency medical services and traffic officials from as far as Rustenburg and Mafikeng joining in to help in making the day a success. Gazing on the Ditrupa performing groups, it can be noted that the tradition has been passed down as a community tradition. Makapanstad as a territory with low population density experiences a huge impact in the form of domestic tourism activity from the Ditrupa Festival. A lot still needs to be done to ensure the locals experience the full benefits of this tourism activity, though some have recognized the opportunities that are brought by the festival. The community of Hammanskraal, which performs the Ditrupa tradition, can use the festival as a means of lobbying for a fully-fledged memorial park and multi-media museum to be constructed to commemorate the participation of Black South Africans in World War I and to document and curate the Ditrupa tradition. This would be a means to create an additional attraction for Hammanskraal. The National Department of Tourism has funding for community tourism projects that are off the beaten track, as a means of empowering the previously disadvantaged groups, and allowing for the geographic spread of tourists. The event could be regarded as a successful heritage event.

Conclusion

Event tourism continues to act as an attraction that draws visitors to a destination, and the cases presented here in this chapter, Berlin's November traditional horse race, the Maletsunyane Braai Festival in Lesotho and the Ditrupa Festival in Hammanskraal, are events that remain small in size and have the potential to grow in attracting tourists and increasing the expenditure in addition to increasing length of stay. The fact that the festivals are yearly events, occurring on designated dates, reflect that they are community-initiated events that have been institutionalized, making them sustainable events. The Berlin November hosted outside East London in the Eastern Cape Province has developed from being a traditional horse racing event to include a fashion parade, and hospitality packages which have increased the amount of investment the event attracts. To ensure the sustainability of the event, more infrastructure and superstructure such as roads and bridges could allow for more private-sector investment in the event. The Ditrupa Festival, which commemorates the heritage of the participation of Black South African soldiers in World War I, is a successful heritage event that occurs during the December recess period. It experiences the active participation of the community, who are on recess and this could be classified as visiting friends and relatives (VFRs) tourism. The Maletsunyane Braai Festival, a barbeque-themed event in the rural hinterlands of Lesotho, is established against the background of the Maletsunyane Falls and was established less than 6 years ago. The event that is themed around camping near the falls and participating in barbeque activities, has diversified the tourist markets to Lesotho. The event had assisted to highlight the Maletsunyane Falls, and has given adventure tourists another destination in Lesotho. Promoting tourism to peripheral areas is imperative to ensure that the developmental impacts of tourism are spread geographically. Spreading the developmental impacts of tourism is imperative to ensure that tourism continues to deliver positive economic impacts to citizens that find themselves located on the urban fringe.

References

Apleni, L. and Henama, U.S. (2020) The impact of events in boosting local economic development: a case study of Port St Johns, South Africa. *African Journal of Hospitality, Tourism and Leisure* 9(1), 1–12.

Clark, C. (2016) They are stealing the sport. Available at: https://www.groundup.org.za/article/theyre-stealing-sport/ (accessed 1 January 2020).

Cycloscope (2019) Maletsunyane Falls in Semonkong, Lesotho: the highest single-drop waterfalls in Africa. Available at: https://cycloscope.net/maletsunyane-falls-semonkong (accessed 1 January 2020).

Etiosa, O. (2012) The impacts of event tourism on host communities. Case: the city of Pietarsaari. Degree programme thesis, Central Ostrobothnia University of Applied Sciences, Pietarsaari, Finland.

Getz, D. (2008) Event tourism: definition, evolution, and research. *Tourism Management* 29(3), 403–428. DOI: 10.1016/j.tourman.2007.07.017.

Giampiccoli, A., Lee, S.S. and Nauright, J. (2013) Destination South Africa: comparing global sports mega-events and recurring localised sports events in South Africa for tourism and economic development. *Current Issues in Tourism* 18(3), 229–248. DOI: 10.1080/13683500.2013.787050.

Henama, U.S. (2017) International migration and national tourism industry development (the case of south africa). *The EUrASEANs: Journal on Global Socio-Economic Dynamics* 2(3), 72–82.

Ledwaba, L.S. (2017) *Cheers to Thundering Hooves of Village Horsemen*. Mukurukuru Media, Johannesburg, South Africa.

Lonely Planet (2015) *South Africa, Lesotho & Swaziland*, 10th edn. Available at: https://www.lonelyplanet.com (accessed 1 January 2020).

Mirele, M.D. (2015) Economic impacts of the FIFA World Cup in developing countries. Honours thesis (Paper 2609), Western Michigan University, Kalamazoo, Michigan.

Oklobdzija, S. (2015) The role of events in tourism development. *Yaconnc* 6(2), 83–97. DOI: 10.5937/BIZINFO1502083O.

SABC News (2018) Diturupa to explore legacy of African World War 1 heroes. South African Broadcasting Corporation (SABC). Available at: https://www.sabcnews.com/sabcnews/diturupa-to-explore-legacy-of-african-world-war-1-heroes/ (accessed 17 March 2020).

SABC News (2019a) Annual Makapanstad Trooper celebrations enhance tourism. South African Broadcasting Corporation (SABC). Available at: https://www.sabcnews.com/sabcnews/scores-of-people-converge-on-makapanstad-for-annual-trooper-event/ (accessed 17 March 2020).

SABC News (2019b) Makapanstad community gathers in song, drums and dance to celebrates diturupa. South African Broadcasting Corporation (SABC). Available at. Available at: https://www.sabcnews.com/sabcnews/makapanstad-community-gathers-in-song-drums-and-dance-to-celebrates-diturupa/ (accessed 1 January 2020).

ShowMe (2019) The Berlin November – Buffalo City's Premier Horse Racing & Lifestyle Event. Available at: https://showme.co.za/east-london/lifestyle/the-berlin-november-buffalo-citys-premier-horse-racing-lifestyle-event/ (accessed 1 January 2020).

Somicimbi, I. (2017) Lesotho, in all its majesty. *News24*, 30 November. Available at: https://www.news24.com/SouthAfrica/Local/City-Vision/lesotho-in-all-its-majesty-20171129 (accessed 1 January 2020).

Strydom, A.J., Saayman, M. and Saayman, A. (2006) The economic impact of the Volksblad Arts Festival. *Acta Commercii* 6(1), 87–98. DOI: 10.4102/ac.v6i1.97.

Umhlobo Wenene (2017) Berlin November. Available at: http://www.umhlobowenenefm.co.za/sabc/home/umhlobowenenefm/events/details?id=b0ed2540-c0b8-4cef-98b3-89db2c313871&title=Berlin%20November (accessed 1 January 2020).

World Economic Forum (2018) Travel & Tourism Competitiveness Report 2018. World Economic Forum, Geneva, Switzerland.

12 Heritage Sporting Events (HSEs): Issues and Prospects

Bayram Akay*
Burdur Mehmet Akif Ersoy University, Burdur, Turkey

Introduction

Sport can be an integral part of a destination's heritage, and invariably some tourists will choose to visit locations associated with a distinct sporting past and sporting culture. Sporting attractions with heritage components, such as stadia, famous sporting locations, halls of fame and sport museums, are often part of a sport tourist's itinerary (Ramshaw and Gammon, 2005). While some destinations were focusing on sports events to brand and promote their territory, other regions were developing strategies based on endogenous resources (local heritage). By focusing on the local identity and the unique features of the territory, these 'heritage-based' strategies can strengthen the economy and attract tourists, while having a positive impact on the quality of life and the attractiveness of the territory (Pinson, 2017).

Major sports events – such as the Olympic Games, the FIFA (Fédération Internationale de Football Association) World Cup, the Tour de France cycling event, Wimbledon Tennis Tournament, the European Rugby Cup, European Swimming Championships, World Athletics Championships and University Games – are the major sports events that attract the most attention today (Toksöz *et al.*, 2017). All sports events are not heritage sports events. A heritage sporting event (HSE) must belong to a specific place and be held there regularly, usually every year, sometimes every 2 years, without interruption other than under exceptional circumstances. The owner of the event must be a local organization, generally a not-for-profit association or a local council, but not an international organization (Chappelet, 2015).

This chapter focuses on the relationship between sports heritage events and tourism. First the historical development of HSEs and the relationship between HSEs and tourism will be considered. This will be followed by examination of the Kırkpınar Oil Wrestling Festival, a sports heritage event identified by the United Nations Educational, Scientific and Cultural Organization (UNESCO) in 2010 when it was added to the List of Intangible Cultural Heritage of Humanity.

Historical Development of HSEs

The tourism event was started by the Sumerians, who first found and used writing, money and the wheel, in the period up to 4000 BC. It can be said that those of Phoenicia were also the first travellers in this sense. Also in 3000 BC, Egypt became one of the countries that attracted the most attention from travellers as people travelled to

*bakay@mehmetakif.edu.tr

 Festival and Event Tourism: Building Resilience and Promoting Sustainability (eds A. Sharma *et al.*)
DOI: 10.1079/9781789248685.0012

the pyramids and temples. In ancient Egypt and the Babylonian Empire, safe roads for travellers, houses and gardens were built for accommodation and relaxation purposes. The Olympics, created by the ancient Greeks, revealed the possibility of travel, as those who went to the Olympics pioneered tourism by staying in towns (Yıldız, 2011).

The Olympic Games, held in honour of Zeus on Mount Olympus in 776 BC, attracted spectators from all over Europe and the Middle East. During and after this period, travel increased as did tourism movements. The Olympic Games have an important place among the steps of the travel and tourism sector; and now it has gone beyond being a sports organization and has become a major tourism activity where vast numbers of national and international visitors travel significant distances and stay for up to weeks at a time (Shackley, 2006; Page, 2007).

Ancient Egypt

The ancient Egyptians were interested in many sports, including wrestling, gymnastic exercises, weight lifting and ball playing. Bullfighting was the most interesting activity, and its beginnings were based on a religious motive. Music, drama and dance were adopted as a form of religious worship as well as social entertainment (McLean *et al.*, 2008).

Ancient Greece

Greek culture was founded mainly on the development of physical education and sports. Two important elements in the training were music and gymnastics. The gymnastics studies applied were also usually made up of studies involving physical activity (Siedentop, 2009, p. 24). A gradual transition occurred in the Greeks' approach to free time and play. At first, all citizens considered participating in sports and games, but the Olympic Games were held exclusively for Greeks. Later, the religious and cultural functions of the Olympic Games and other festivals were weakened due to the prominence of both malicious people and commercial attitudes. Other forms of activity such as drama, singing and dancing with sports began to be implemented by talented people (usually people selected from the low class or slave class) who trained for many years before appearing before the audience and focused only on their skills (McLean *et al.*, 2008, pp. 53–54).

Ancient Rome

The Romans supported the Olympic Games not for aesthetic and religious purposes, but for their benefit. The Romans were less interested in forms of cultural activity than the Athenians. In addition, the Romans seemed to be very systematic planners and were a creative civilization compared with the Greeks. In Roman towns, bath areas, open-air theatres, sports fields, forums that brought the public together, stadiums, parks and gardens were quite common. The Romans built facilities for gyms, and the wrestling rooms were modelled on Greek gymnasiums, which included interview/chat spaces for philosophers. Baths, amphitheatres and arenas were built for the benefit of all the public. In Rome, there were more than 800 public baths, and they were free or cost little to use, but free-time activities were sometimes inadequate. Emperors tried to please the public by providing free food and entertainment. Also, slaves were used not only as workers, but also for entertainment purposes such as music, drama and sports (Torkildsen, 1999; Bhatia, 2002).

Ancient Assyria and the Babylonian Empire

The Assyrian and Babylon civilizations are marked by having very popular recreation activities including boxing, wrestling, archery and different board games. In addition to watching dancing, listening to music and feasting activities, hunting activities were among the most common activities for the Assyrians. The nobles of the Assyrians used to go hunting for lions in wheeled cars or on foot and hunted with spears. Hunting was seen as a daily pursuit, and there are many reliefs, sculptures and inscriptions depicting it (McLean *et al.*, 2008).

Industrial Revolution (1775–1900) and later

The Industrial Revolution brought about a decrease in people's leisure time, but by the 20th century, an increase in leisure time was seen again. In the 1920s, hotels, holiday services and train, ship, automobile and bus travel began to become widespread (McComb, 2008).

On 23 June 1894 it was decided that the Modern Olympic Games would be held every 4 years and the first was held in Athens in 1896, and the International Olympic Committee (IOC) was established (Yıldıran, 2014, p. 565). However, there have been periods when the Olympic Games could not be held, for example in 1916 due to World War I, and in 1940 and 1944 because of World War II (Şentürk and Özdilek, 2007). In 2020, national and international sporting events such as the Olympic Games (Tokyo), the European Football Championships, the French Open Tennis Tournament, Formula 1, NFL (National Football League), baseball and hockey, and the Kırkpınar Oil Wrestling Festival were postponed due to the coronavirus 2019 (COVID-19) pandemic.

HSEs and Tourism

Sport can act as an important legacy for individuals, regions and nations. Perhaps few other links to the past are as passionately expressed as those for sport and its heritage. Fans identify with the past of their team, even if that past does not exist in their living memory, while others see sport as the continuity of a national, regional or personal legacy and identity (Ramshaw and Gammon, 2005). Sports heritage consists of four main elements: (i) tangible immovable property (places, spaces connected to the past); (ii) concrete portable items (objects, concrete experiences); (iii) intangible items (rituals, traditions, chants, memories, other items not held by the hand); and (iv) goods and services with a sport heritage component (tourism, marketing, management) (Korkmaz, 2019).

Sport events provide long-standing benefits such as new infrastructure and facilities, a city's reconstruction, international credibility, tourism growth, public welfare improvement, job creation and local employment opportunities (Honari *et al.*, 2010). Over the last 20 years, many cities, regions and countries have introduced strategies and policies to systematically host sports events, and have allocated human and financial resources to attract elite or mass-participation single-sport or multisport competitions (Chappelet, 2015).

The Olympic Games, Formula 1, the FIFA World Cup, the Tour de France, Wimbledon Tennis Tournament, the European Rugby Cup, the European Swimming Championships, the World Athletics Championships, the UEFA (Union of European Football Associations) Champions League and University Games are the most important sporting events today. Not all sports events are heritage sports events. As stated earlier, an HSE must belong to a specific place and be held there regularly, usually every year, sometimes every 2 years, without interruption other than under exceptional circumstances. The owner of the event must be a local organization, generally a not-for-profit association or a local council, but not an international organization (Chappelet, 2015).

Some sporting disciplines have been assigned to the List of Intangible Cultural Heritage of Humanity by UNESCO, such as equitation in the French tradition (a school of horseback riding that emphasizes harmonious relations between humans and horses). However, only three *events* (that can be seen as sports events rather than sports disciplines) have been assigned so far: (i) the Kırkpınar Oil Wrestling Festival in Turkey; (ii) the Naadam in Mongolia; and (iii) the Dragon Boat Festival in China. The Kırkpınar's form highlights the cazgir's prayers (the cazgir is the master of the festival) and the participants' traditional costumes, while the Naadam and the Dragon Boat Festival focus on the memorial ceremonies that take place during the events (Pinson, 2017).

Issues and Prospects of Events

Sports, business, entertainment and cultural events have some favourable economic, environmental and sociocultural effects (Getz and Page, 2016). However, they can also create some unfavourable effects in the destination where the event is held such as increasing the cost of living, increasing traffic density, crowds, noise, crime and damage to property (Yolal, 2019a). Sharma *et al.* (2008) noted some important problems caused by these events as follows: (i) traffic problems; (ii) environmental and noise pollution; (iii) crowds and turmoil; (iv) damage to some important cultural and historical buildings; and (v) excessive busyness in social services. Events usually take place in the summer months in most destinations. This poses a challenge for the destination management organizations (DMOs), and problems such as congestion, pollution, noise and accidents affect tourists and locals alike. Holding events during the off-peak tourism season is also important for DMOs, tourists and locals in terms of participation. Other issues associated with events are as follows (Kruger, 2010; Getz and Page, 2016; Zhang *et al.*, 2019, p. 2):

- commodification of heritage;
- sustaining cultural and natural heritage;
- environmental and green issues;
- seasonality of demand;
- the impact of climate change and the need for venues and event stagers to consider the capacity to adapt to and mitigate the impacts;
- security and safety issues;
- idle investments in infrastructure and superstructure; and
- globalization and the global audiences for event tourism.

Conferences, trade shows, concerts, sports events and festivals are examples of important events in the event industry (Çobanoglu *et al.*, 2021, p. 54). The great majority of events are sports or cultural festivals organized by the public sector, local authorities and non-profit organizations. These events build up various expectations in all stakeholders (local and central government, the public, sponsors, performers, organizers, businesses, suppliers, volunteers, employees, visitors, media), especially in a particular city, region or country (Andersson and Getz, 2008).

Governments (official institutions) organize events in order to gain some social, cultural, touristic and economic benefits (e.g. additional income, tax, employment) (Kruger, 2010). On the one hand, local authorities expect to ensure that locals take part in the events, a better performance is provided to the public, events contribute to the local economy, and the principle of environmental awareness is followed while organizing events. Employees want to work in a guaranteed job with good pay and career prospects in a safe and comfortable environment. On the other hand, locals want the construction of new roads and facilities in order to have an increase in their quality of life, to socialize, brand the city, and meet their economic expectations. Events such as festivals, congresses and carnivals are also the most popular activities in terms of sponsorship. Sponsors want the sports events to be widely covered in the media so that they can reap the benefits of promoting their organization. The media expects the events to be promoted to a large mass of people. Visitors/tourists participating in the events want to have cultural, artistic, sports and religious experiences (Yolal, 2019b). All businesses such as hotels, restaurants, travel agencies and markets operating in the host city/country focus on events (Presenza and Iocca, 2012).

Research Methodology

The purpose of this research is to assess the contribution to tourism of HSEs by examining the Kırkpınar Oil Wrestling Festival, which has HSE features, as a case study. The research uses a qualitative research method of reviewing the literature, in particular by obtaining data from scanning all written and verbal records kept retrospectively such as books, newspapers, magazines, reports, encyclopaedias and letters (Yıldırım and Şimşek, 2008, p. 8). Quantitative and qualitative content analysis can be made on texts and documents (Altunışık *et al.*, 2005, p. 260).

The Case of Kırkpınar Oil Wrestling Festival in Edirne Province, Turkey

The heritage sport of oil wrestling (*yağlı güreş*) is a particular form of freestyle wrestling performed on unmown grass, with the protagonists and their trousers completely drenched in olive oil, turning victory over the opponent into an extremely slippery and demanding task (Krawietz, 2018). Kırkpınar, which has been held since 1361, the date of the conquest of Edirne, has a history of 700 years. The winner of traditional wrestling is given the title 'Başpehlivan' ('Top Wrestler'). Most of the pehlivans (wrestlers) grown in the meadows of Kırkpınar represent Turkey in Greco-Roman wrestling all over the world. In 2010, the Kırkpınar Oil Wrestling Festival, an annual event, was recognized by UNESCO when it was added to the List of Intangible Cultural Heritage of Humanity (Edirne Provincial Directorate of Culture and Tourism, 2020a).

Kırkpınar oil wrestling is a cultural event with its elements, rituals and cultural identity. The week of wrestling, which is visited by many local and foreign people, is referred to as 'Kırkpınar Week'. While wrestling competitions are held in the last 3 days of this week, cultural events are held in the other time period (Küçükaltan *et al.*, 2005). Likewise, in Turkey every year about 120 oil wrestling matches are held in 31 provinces of Turkey and the number of spectators in a season is about 800,000 people (Özdemir and Çalışkan, 2018). Over 100,000 visitors came to Edirne for Kırkpınar oil wrestling between 9 and 15 July 2018. Hotels and restaurants were 100% occupied, and large amounts of city-specific souvenirs were sold (Hacıoğlu, 2018).

Wrestling, which is considered as an ancestral sport in Turkey, is one of the important sports in the country that attract tourists both from the Turkish community and from elsewhere in the world. The most important feature of Kırkpınar is that it has continued for centuries without losing its traditional characteristics (Soygüden, 2018). Oil pehlivan wrestling is also held in other provinces and districts of Turkey. For example, as stated above, in the 2018 season, the number of oil wrestling organizations held on various scales reached 120. The wrestling season covers the period between April and October (Özdemir and Çalışkan, 2018). On 5–7 July 2019, the 658th Historical Kırkpınar Oil Wrestling event was held in the Private Square wrestling area of Edirne Palace. A total of 2381 athletes participated in the competition in all categories (Soygüden *et al.*, 2020). In 2020, the 659th Kırkpınar Oil Wrestling event, due to be held between 29 June and 5 July, had to be postponed until 2021 due to the COVID-19 pandemic.

An Overview of Kırkpınar Oil Wrestling Concepts

The name of Kırkpınar oil wrestling

Kırkpınar is described as the gushing of 40 springs from the place where the wrestlers named Ali and Selim from the 40 brave commanded by Orhan Bey's on Süleyman Pasha were martyred and buried. It is known as 'Rumelia wrestling' among the local people. Kırkpınar is defined as a form of wrestling based on a competition, which includes oil wrestling, religious, national rituals and Turkish customs and traditions, and examples of charity and hospitality (Sengül, 2020, p. 600).

Elements and rituals of Kırkpınar oil wrestling

The word 'ritual' means a religious or solemn ceremony consisting of a series of actions performed according to rules. Ritual is defined as all kinds of traditional and formal practices created by those who have a belief by putting symbolic meaning into an alliance (Sengül, 2020, p. 601). The elements and rituals of Kırkpınar are as follows.

1. **Wrestling games:** According to the rules of oil wrestling the loser is considered to be the wrestler: (i) whose back comes to the ground (showing their belly to the stars) as a result of the opponent's actions; (ii) who sits with support on two hands behind; or (iii) who touches the ground with two elbows or an elbow and a hand (Bakhrevskiy, 2019). There are many techniques applied by wrestlers to outperform their opponent in competitions including, for example, *ters kabza* (back handle), *iç kabza* (inner handle), *dış kabza* (outer handle) and *göğüs şaki* (chest grabbing) (Edirne Provincial Directorate of Culture and Tourism, 2020b).

2. **Peshrev:** Peshrev is a combination of a cultural ritual and physical action, which takes place before wrestling starts. To musical accompaniment the pehlivans perform a variety of movements in time with the various rhythms. These harmonious movements entertain the audience and raise the spirits of the pehlivan. The pehlivan prepares his breath, muscles and heart for the challenge that begins in a little while.
3. **Oiling:** In order to make the wrestlers difficult to grasp, each pehlivan is covered in lubricant; this takes place around cauldrons filled with oil and water in a convenient part of the wrestling square. After wrestling begins, pehlivans can get oil and water from carriers walking around the meadow at any time (Edirne Provincial Directorate of Culture and Tourism, 2020a).
4. **Kispet:** The only unchanging garment of oil wrestling that has survived for centuries is the kispet; the creativity of handicrafts is one of the most important elements of wrestling. The kispet is a garment that covers the body from the wrestler's navel to below the knees. The waist part of the kispet is made of buffalo and young calf skin and is four fingers wide and thick. Thick rope is passed to wrap the back; this part is called the 'kasnak'. The place under the knee cap of the kispet is called the 'paça'. Between the leg and the calf, a felt called the 'paçabent' is placed; the leather part is pulled over the felt and tightly tied with the rope (Kaygusuz *et al.*, 2019).
5. **Prayer:** As a living carrier of verbal cultural heritage, jazz lists provide a warm dialogue between people from different cultures with their literary poetic presentations called 'prayer'. Prayer, which is one of the important rituals of Kırkpınar oil wrestling, are the lyrics (sung by the jazz musicians) that are derived from the words that the cazgir spoke in verse before he released the wrestlers to wrestle. Jazz musicians continue this tradition from the past to the present (Sengül, 2020).
6. **Red-bottomed candle:** The invitation symbol of Kırkpınar is a red-bottomed candle. In times when mass media was not common, the red-bottomed candle was sent to the towns and villages, to wrestlers and well-known people by March, and they were invited to Kırkpınar. The phrase 'calling with a red-bottomed candle', which is used frequently among the people, comes from this. Today it is used as a symbol. City shopkeepers are also invited to the Kırkpınar Oil Wrestling Festival with a red-bottomed candle, accompanied by drum and zurna (Sengül, 2020).
7. **Zembil:** The zembil is a traditional straw bag made by hand on a special loom (Özdemir and Çalışkan, 2018). Wrestlers put their wrestling equipment (e.g. the kispet) in the zembil.
8. **Golden belt:** The golden belt was prepared for the first time by Edirne Municipality in 1969 as the biggest award for wrestling. The belt is decorated with moon and star motifs on the right and left of the reliefs of Kurtdereli Mehmet Pehlivan and Adali Halil Pehlivan in the middle of the golden belt. To be awarded the 'golden belt', a wrestler must be the champion three times sequentially at the Kırkpınar Oil Wrestling Festival. The champions in Kırkpınar oil wrestling are called the 'Top Wrestler' (Bayraktar, 2017, p. 45).
9. **Awards:** Prize money for the tournament is given by Edirne Municipality. The amount of the award is determined according to the competition categories.
10. **Agha of the Kırkpınar Oil Wrestling Festival:** In the past, wrestling was held under the control of the agha, its patron. He is the authority that calls the wrestlers, organizes the competitions, hosts the guests, provides food and accommodation, ensures wrestling takes place in accordance with customs and traditions, and gives the awards (Edirne Provincial Directorate of Culture and Tourism, 2020b).

Conclusion

Sports events can have positive effects (e.g. on sports infrastructure, urban transformation, transportation, accommodation, telecommunications, image promotion and employment) and negative effects (e.g. maintenance costs and the inert capacity of sports facilities that cannot be used during the event, environmental degradation) on country/city tourism (Çetin, 2019). Many of the problems experienced by hosts arise after sporting events. Therefore, cities that host sporting events should take the elements of the sports heritage concept with a holistic approach and minimize the problems that may arise after the end of the event.

Sport can act as an important legacy for individuals, regions and nations. Activities with sports heritage elements such as Kırkpınar oil wrestling should be supported. Mega-events should be held by the state, as they do not reflect both investment costs and sports heritage. For example, wrestling is an event at the regional level with elements of sports heritage from past to present. On the 3–4 days of the historical Kırkpınar Oil Wrestling Festival, spectators from all over Turkey come to Edirne. These activities consist of concrete values that convey information to us about how our ancestors lived and the activities they did. What is different from other bodily activities of traditional sports is that they are in the form of recreational games and they gain prevalence by being accepted by the public and passed on through generations (Soygüden, 2018).

Different destinations in many countries are realizing that the event industry is vital to attract visitors and they are trying to create new events. Events also help to prolong the season and reduce the seasonality effect in tourism, especially in destinations with seasonal demand. For example, in the tourism centre of Ölüdeniz, in the Fethiye district of Muğla province, the International Fethiye Ölüdeniz Air Games Festival has been held for 20 years to extend the summer season and promote the region. During the aerial games, hot air ballooning, paragliding acrobatics, skydiving, powered paragliding, base jumping, microlight and model aircraft shows are held.

References

Altunışık, R., Coşkun, R., Bayraktaroğlu, S. and Yıldırım, E. (2005) *Research Methods in Social Sciences (SPSS Applied)*, 4th edn. Sakarya Publishing, Sakarya, Turkey.

Andersson, T.D. and Getz, D. (2008) Stakeholder management strategies of festivals. *Journal of Convention & Event Tourism* 9(3), 199–220. DOI: 10.1080/15470140802323801.

Bakhrevskiy, E. (2019) History and actual image of oil wrestling. *International Journal of Ethnosport and Traditional Games* 2(2), 12–36.

Bayraktar, K. (2017) An oil wrestler in the game world: hakan. *Journal of Edirne Kırkpınar* 3, 40–45.

Bhatia, A.K. (2002) *Tourism Development: Principles, Practices*. Sterling Publishers, New Delhi.

Çetin, A. (2019) Necessity of the legacy concept in the mega sport event. *Journal of Health and Sport Sciences* 2(2), 66–73.

Chappelet, J.L. (2015) Heritage sporting events and place marketing. In: Plevnik, M., Retar, I., Pišot, R. and Obid, A. (eds) *Sustainable Development of Sport Tourism*. Annales University Press, Koper, Slovenia, pp. 57–72.

Çobanoglu, C., Doğan, S. and Güngör, M.Y. (2021) The integration of social media into event tourism. In: Unurlu, Ç. (ed.) *Emerging Technologies at Events*. IGI Global, Hershey, Pennsylvania, pp. 53–68.

Edirne Provincial Directorate of Culture and Tourism (2020a) Tanıtım Materyalleri. Available at: https://edirne.ktb.gov.tr/TR-76489/tanitim-materyalleri.html (accessed 20 January 2021).

Edirne Provincial Directorate of Culture and Tourism (2020b) Kırkpınar Oil Wrestling. Available at: https://edirne.ktb.gov.tr/TR-76392/Kırkpınar-yagli-guresleri.html (accessed 20 January 2021).

Getz, D. and Page, S.J. (2016) Progress and prospects for event tourism research. *Tourism Management* 52, 593–631. DOI: 10.1016/j.tourman.2015.03.007.

Hacıoğlu, A. (2018) Kırkpınar was productive in terms of tourism. Available at: http://www.edirneyenigun.com/haber/19824/Kırkpınar-turizm-acisindan-verimli-gecti.html (accessed 10 January 2021).

Honari, H., Goudarzi, M., Heidaric, A. and Emamid, A. (2010) A comparison of the viewpoints of tourists, interested managers and cultural heritage organization managers regarding sport tourism-driven job and income creation in Mazandaran, Iran. *Procedia Social and Behavioral Sciences* 2(1), 5659–5663. DOI: 10.1016/j.sbspro.2010.03.924.

Kaygusuz, M., Sarı, H.K. and Arğun, F.N. (2019) 'Kispet' – traditional clothing of the cultural heritage oil wrestling sport. *Akademik Bakış Dergisi* 72(1), 147–166.

Korkmaz, S. (2019) Heritage transfer in sport culture: the demolition example of Ankara 19 Mayıs Stadium. *SPORMETRE* 17(3), 9–22. DOI: 10.33689/spormetre.470153.

Krawietz, B. (2018) Ways of knowing Muslim cultures and societies. In: Gräf, B., Krawietz, B., Amir-Moazami, S., Freitag, U. and Hirschler, K. (eds) *On Coming to Grips with Turkish Oil Wrestling: Conceptualising Muscular Islam and Islamic Martial Arts*. Koninklijke Brill NV, Leiden, the Netherlands, pp. 327–354. DOI: 10.1163/9789004386891.

Kruger, E.A. (2010) A stakeholder perspective on mega-events as an element of tourism destination competitiveness. Dissertation for MCom Degree in Tourism Management. University of Pretoria, Pretoria, South Africa.

Küçükaltan, D., Oğuzhan, A., Apak, S. and Boyacıoğlu, E.Z. (2005) The impact of cultural tourism on regional development: a sample of Kırkpınar wrestling. *Trakya University Journal of Social Science* 6(1), 1–22.

McComb, D.G. (2008) *Spare Time in Texas: Recreation and History in the Lone Star State*. University of Texas Press, Austin, Texas. DOI: 10.7560/718708.

McLean, D.D., Hurd, A.R. and Rogers, N.B. (2008) *Kraus' Recreation and Leisure in Modern Society*, 9th edn. Jones and Bartlett Publishers, Burlington, Massachusetts.

Özdemir, G. and Çalışkan, V. (2018) Geography of a traditional sport: oil wrestles in Turkey (wrestles in Kırkpınar, Edirne and Çardak, Çanakkale). International Geography Symposium on the 30th Anniversary of TUCAUM, 3–6 Ekim 2018/3–6 October 2018, Ankara.

Page, S.J. (2007) *Tourism Management: Managing for Change*. Elsevier Butterworth-Heinemann, Oxford, UK.

Pinson, J. (2017) Heritage sporting events: theoretical development and configurations. *Journal of Sport & Tourism* 21(2), 133–152. DOI: 10.1080/14775085.2016.1263578.

Presenza, A. and Iocca, S. (2012) The weight of stakeholders on festival management. The case of music festivals in Italy. *Revista de Turismo Patrimonio Cultural* 10(2), 25–35. DOI: 10.25145/j.pasos.2012.10.024.

Ramshaw, G. and Gammon, S. (2005) More than just nostalgia? Exploring the heritage/sport tourism nexus. *Journal of Sport Tourism* 10(4), 229–241. DOI: 10.1080/14775080600805416.

Sengül, D. (2020) 'Concepts and rituals' Kırkpınar Oily Wrestling Festival as UNESCO unsute cultural heritage. *Research & Investigations in Sports Medicine* 7(2), 600–606. DOI: 10.31031/RISM.2020.07.000657.

Şentürk, A. and Özdilek, Ç. (2007) Reasons of burning the olympiade fire in antique and modern olympiade games and conduction the flambeau in the games that organized until present. *Dumlupınar University Journal of Social Sciences* 18, 217–230.

Shackley, M. (2006) *Atlas of Travel and Tourism Development*. Elsevier Butterworth-Heinemann, Oxford, UK.

Sharma, B., Dyer, P., Carter, J. and Gursoy, D. (2008) Exploring residents' perceptions of the social impacts of tourism on the Sunshine Coast. *International Journal of Hospitality & Tourism Administration* 9(3), 288–311. DOI: 10.1080/15256480802096092.

Siedentop, D. (2009) *Introduction to Physical Education, Fitness, and Sport*. McGraw-Hill, New York.

Soygüden, A. (2018) Kırkpınar oil wrestling, spectator, social-demographic features and reason of participation. *Trakya University Journal of Social Science* 20(2), 337–352.

Soygüden, A., Toksöz, İ. and Özen, T. (2020) 658. Historical Kırkpınar oil wrestling; examination of spectators opinion of oil wrestling techniques and head oil wrestler. *Akademi Sosyal Bilimler Dergisi* 7(21), 60–75. DOI: 10.34189/asbd.7.21.004.

Toksöz, D., Bak, E. and Sercan, B. (2017) Mega ve özellikli spor etkinliklerinin dünyadan ve Türkiye'den örneklerle benzerlik ve farklılıkları. *3. Doğu Akdeniz Turizm Sempozyumu* 18 Nisan 2014. Mersin.

Torkildsen, G. (1999) *Leisure and Recreation Management*, 4th edn. Routledge, London.

Yıldıran, İ. (2014) Antikiteden modemiteye olimpiyat oyunları: idealler ve gerçekler. *Hece (Batı Medeniyeti Özel Sayısı)* 18(210–212), 555–570.

Yıldırım, A. and Şimşek, H. (2008) *Qualitative Research Methods in Social Sciences*, 7th edn. Seçkin Publishing, Ankara.

Yıldız, Z. (2011) Development of tourism sector and its effects on employment. *Suleyman Demirel University, The Journal of Visionary* 3(5), 54–71.

Yolal, M. (2019a) Kongre ve etkinlik yönetimi. In: Özel, Ç.H. and Sezerel, H. (eds) *Etkinliklerin Sosyo-Ekonomik Ve Kültürel Etkileri*. Anadolu University, Eskişehir, Turkey, pp. 66–87.

Yolal, M. (2019b) Kongre ve etkinlik yönetimi. In: Özel, Ç.H. and Sezerel, H. (eds) *Etkinlikler Ve Paydaşlar*. Anadolu University, Eskişehir, Turkey, pp. 112–133.

Zhang, T., Wen, H. and Li, X. (2019) A tourist-based model of authenticity of heritage sporting events: the case of Naadam. *Sustainability* 11(1), 108. DOI: 10.3390/su11010108.

Part 4: Case Studies on Festival and Event Tourism Destinations: Branding and Promotions

13 Women, Religion and Festivals: Exploring Qualitative Dimensions of the Role of Women in Legends Behind the Celebration of Festivals in India

Manpreet Arora*
School of Commerce and Management Studies, Central University of Himachal Pradesh, Dharamshala, India

Introduction

Celebrating life through festivals in India is a way of life. Celebrating life is taken as being synonymous with festivals in India as they are actually a way of living life to the fullest. There are a lot of festivals and fiestas which are celebrated throughout the year in various religions. They reflect the cultural heritage of the country and the diversity of rich traditions across the state. Every occasion throughout the year calls for celebration in Indian lands whether it is spring, winter, the rainy season, harvesting of crops or the beginning of the year; there are several reasons to celebrate every occasion in the form of festivals. All these festivals bring an extra dose of colour in the form of decorations, clothing, foods, music and dance. Many festivals are about celebrating the existence of God, divas and deities.

Just as there is a close connection between religion and festivals, women and festivals are also inseparable in India. Almost every religion that is celebrated here highlights the role of women in the form of certain festivals, customary rituals, traditional fairs or even ceremonies performed in each household. Each year, there is a range of festivals which are celebrated with great pomp and show involving the community that belongs to different religions. Amid such razzamatazz of rites, rituals and religious ceremonies, women by and large remain pivotal. Such festivals/rituals are sometimes celebrated/observed through simple fasting, and at other times, a lot of money is splurged on the celebrations/observance, which boosts the economy considerably and provides livelihoods to many. Some of the religions tend to celebrate and glorify Mother Nature, which symbolizes womanhood. Interestingly, there are some festivals where the core focus remains on women but sometimes the celebration seems to favour a male god or men. Nevertheless, we cannot imagine celebrating any of the festivals without according women a place of centrality. Say, for example, during *Diwali* (the most popular festival of India), the bulk of activities revolve around women, and these range from shopping, to performing rituals, cooking delicacies, decorating houses, attending to the guests and visitors and arranging family gatherings. Moreover, in the evening on *Diwali*, the goddess *Lakshmi* is worshipped in order to bring prosperity to the family. The whole house

*arora.manpreet3@hpcu.ac.in

DOI: 10.1079/9781789248685.0013

is decorated with lights and lamps and colourful floor designs called *rangoli*, as it is believed that goddess *Lakshmi* will arrive at every home to bless the worshippers with health, wealth and prosperity. No one in a Hindu family can ever think of celebrating *Diwali* without the presence of women in one form or the other.

The author believes that women are the soul of most of the festivals/rituals in Indian tradition. However, it is a country where women suffer as they are deprived of their rights at various levels. Many people in the past had to struggle for equality of men and women in Indian society. Even now there are many places where women are not safe but this is changing gradually. The degree of freedom which women enjoy in India is increasing too but still there are women who are not allowed to study and remain confined to the four walls of the house. Crime against women and rape cases are on the rise.

Since time immemorial in India, women have exemplified their courage and valour in various arenas. Festivals are treated as the mirrors of every society wherein the religious, social and cultural lives of people are reflected (Sharma, 2008; Mair, 2018). If we try to assess the situation of women in the Indian scenario, we should have a look at why and how these festivals are celebrated. What are the myths/realities/backgrounds/conceptions behind these celebrations from the viewpoint of women? In the Indian scenario, women and festivals remain inseparable (Wadley, 1977, pp. 113–125.). Especially in Hinduism, the religious life is closely knit with celebrating various ceremonies, festivals and rituals. Almost every festival has got a certain religious background and women are an integral part of it (Raman, 2009).

In this chapter the author attempts to assess the main religious events of India in view of their traditional and sociocultural importance to ascertain the role women play during such celebrations, and also to underscore how their importance is otherwise relegated. The idea is to lend prominence to the difference between appearance and reality vis-à-vis women's actual importance and their willful and manipulative relegation to an obscure position by the patriarchal mindset. Such a scenario will be critically analysed by taking recourse to the available literature, news articles and other online secondary sources available.

Methodology

This chapter seeks to employ qualitative methodology to conceptually argue that just as there is a close connection between religion and festivals, women and festivals are also inseparable. The study seeks to examine how germane women are in the very conception as well as the celebration of the most prominent festivals in India, and also to demonstrate how women were always respected and kept on a high pedestal in the Hindu legends that are behind the celebration of many festivals. In the course of developing the chapter, the qualitative data analysis software NVivo (QSR International, nd) was used for social media analysis to assess public sentiment regarding the perception of the importance and role of women in Indian festivals, rituals and ceremonies – an area that requires contestation at various levels to redefine, reassess and repurpose the role of women.

History of the Position of Women in India

Most festivals in India are celebrated on auspicious days as per the religious calendars and they are celebrated to mark something in particular. If one wants to understand India or Indian people then one must understand and know about Indian festivals. All of them are quite colourful and they are full of life and enthusiasm. Most festivals are religious and women are an inseparable part of these religious ceremonies both in their preparation and during the festivals. From early times, especially in ancient India, the role of women was regarded with great honour and a lot of the responsibilities of a family lie with them even now. However, in later periods they have not been treated with the honour and respect that was mentioned in earlier scriptures/sagas/folklore, etc. Actually a lot of changes can be seen in the status of women over the period of time. The position of women deteriorated especially in the Indo-Aryan period and continues to remain abysmal in the modern period. Many reformers worked hard to uplift the status of women in India during the British Raj. Now women can be seen working in every area, in higher positions and their recognition is increasing in every walk of life. The author wants to highlight the fact that in true terms the ancient Vedic culture of

India placed women in a very respectable position which in later periods deteriorated. There are many reasons for the deterioration in the status of women as India was invaded by many rulers. But the real essence of the status of women in Indian traditions can be judged through its fairs, festivals, folklores and rituals, prevalent from ancient times, which convince that women were considered the source of life and power and had a much more respectable position than men. Even now, they are actually the pillar of every family.

Further, women are often at the *centre* of festivals. For example, in the Himalayan region interestingly there is a *deity* or *devta* (local god and goddess) for every region. The diversity in Indian festivals can be seen in many ways and is more culturally diverse than can be imagined; even the food, culture, religion and music is widely diverse. In some parts of Odisha and Tamil Nadu in India the menstruation of women is celebrated. It signifies the mark of the transition of when a girl becomes a woman; it is regarded as when a woman becomes fertile, as in nature. In Odisha, a festival named *Raja Prabha* is organized by women to celebrate menstruation. In Tamil Nadu there is a bathing ceremony organized by women for 9 days to celebrate womanhood.

Although there are many instances where women have been marginalized in global history, the author contends that the Indian tradition has always been to take women as an integral part of life. Some of the Indian festivals are described in the next section with the legends behind these celebrations.

Festivals – the Indian Way

Diwali

One of the most spectacular functions where the centrality of a woman is the basis of all celebrations is *Diwali*. This is the festival that involves decorations, lights, food and fireworks. Can anyone in India think of celebrating without giving honour to women? The answer is a 'No'. The reality is that the society which undermines the position of women at various levels, in which a distinction is made between a man and a woman on the basis of their earning capacity, is the society that worships and wants to please *Devi Lakshmi* (goddess *Laxmi*) for wealth and prosperity, and of course she is a woman. People decorate their homes and clean their homes so intensely before *Diwali* in the belief that *Devi* will visit everyone's home and if the homes are clean, decorated and beautiful she will stay at that home giving immense wealth and prosperity to the households.

Diwali is basically a 5-day long festival and is celebrated 20 days after the *Vijyadashmi* (i.e. *Dashehra*). *Diwali* or *Deepawali* originates from the Sanskrit words *deep* (lamp) and *vali* (row). The literal meaning of *Diwali* is 'row of lights'. Earthen lamps are lit by people in homes and temples in the evening. *Diwali* is celebrated not only by Hindus but also to mark different events by different communities including Jains, Sikhs and Buddhists. It basically symbolizes the victory of spirituality (i.e. the victory of light over darkness) which in turn symbolizes the victory of good over evil as well as the victory of knowledge over ignorance. One popular legend associated with the celebration of *Diwali* is that goddess *Lakshmi* was born through '*Samudra Mantham*' (i.e. the churning of the 'cosmic ocean of milk by gods and demons'). On the night of *Diwali*, the goddess chose her husband *Vishnu* and married him. *Lakshmi* not only signifies wealth and prosperity, it signifies much more beyond in terms of knowledge, skills and talents. *Lakshmi* is regarded as a cosmic spiritual energy that manifests as the entire spiritual and material well-being of a person.

The first day of the festival is regarded as *Dhanteras*, where women clean their homes and make colourful decorations on the floor known as *rangoli*. People shop for gold or kitchen utensils to help bring good fortune. On the second day it is *Naraka Chaturdashi*, or *Kali Chandes*. This day is also associated with overpowering the dark evil. In this festival women play a central role. The mere form of *Kali* represents power, feminine energy and to an extent violence too. She is full of creativity and symbolizes fertility. She is considered to be the master of death, change and time. In various respects *Kali* appears to be frightening and in a battlefield killing demons. But she is regarded as the ultimate mother figure. She is a protector and liberator. In another way she is depicted as a mother who is fierce, and is very protective of her children. She is in fact a power who can rise to kill anyone who threatens her children. The *Hindu* devotion towards *Kali* has various aspects but whatever it is she is a feminine aspect of our own existence. The word *Kali* comes from the Sanskrit word *Kal* that means time. So she is a signifier that there is nothing that escapes time. She is also associated with liberation and incarnation. So many spiritual practices are associated with her where she ends the illusion of life and ego. The third day of the festival is the day of *Lakshmi Puja* and is the darkest night of the traditional month.

Various holy hymns, writings and *shlokas* (verses or poems) in ancient scriptures describe the gods and goddesses. Among one of them is *Ashta Lakshmi* in which some lyrics are recited by devotees to accomplish their wishes from the goddess. The goddess, who is a woman, is regarded as a giver of almost everything. The excerpt from *Ashta Lakshmi Stotram* with the translation below is reproduced here and illustrates the central place of women in life.

> *Sumanasa Vandita Sundari Madhavi Chandra sahodari hemamaye*
>
> *Munigana venditha moksha-pradayanini Manjula bhashini vedanuthe*
>
> *Pankaja-vasini Deva supujitha Sadguna varshini santhiyuthe*
>
> *Jaya Jaya hae Madhusudhana kamini Adi Lakshmi sada palaya mam*
>
> [**Translation**]
>
> O the Goddess Aadi Lakshmi, the one who is worshipped by the good, the enchanting one, the consort of Madhava and the sister of Chandra, the one filled with the gold
>
> O Goddess, who is praised by the Muni's and their followers, the bestower of Moksha (salvation), whose speeches generates happiness, the Goddess praised by the Vedas
>
> The one residing in the Lotus, worshipped by the Devas, the giver of good qualities, the one with a pleasant heart
>
> Victory, Victory to you, the dear (consort) of Madhusudhana, Goddess Adi Lakshmi, please rule me with your mercy.
>
> (Hinduism Outlook, nd)

This is one *shloka* from the hymn where the female is worshipped for happiness, which is praised by all and is a giver of salvation. India is a land of gods as well as equally powerful and much more respected women and this has been the case since the ancient period; as such we need to rethink and bring about respect for women in our daily lives.

So the festival of *Diwali* is associated with goddess *Lakshmi*, the goddess of prosperity and wealth, and in some places in the same festival goddess *Kali*, believed to symbolize victory of good over evil, is also worshipped. The general view about goddess *Lakshmi* is that she is the one who grants wealth to devotees but she represents it in a much wider way and includes things like fortune, health, family and overall well-being. The point is the overall well-being of a devotee is addressed towards *shakti* 'the womanhood', so how is it that the patriarchy comes over and undermines even the existence of women who are regarded as the givers of every good thing and fortune of life? The author strongly believes in the power of women to be the backbone of the entire society as signified by deities and goddesses such as *Lakshmi* and *Saraswati*. The value system of India considers women to be the source of existence and granter of all the indispensable things of life. However, a phase came where the value of women in society was undermined and various social evils crept up to exploit women. Therefore in this scenario we need to revisit, rethink and revalue the role and status of women in contemporary Indian society.

Navratri

Navratri is one of the most important and prominent Hindu festivals celebrated in India to worship the power of women and it is done through worshipping the nine divine *avatars* of the goddess *Durga*. The festival spans nine divine nights which are dedicated to goddess *Durga* which is the symbol of *Shakti* 'the power' and spirit of women, female strength and courage as well as the core of the entire existence of the universe. The author contends that it is one of the most significant as well as culturally rich festivals which not only celebrates womanhood but also reminds us that the whole existence of the universe is dependent upon a woman. *Navratri* has been derived from the Sanskrit words *Nava* and *Ratri* which means nine nights.

The legend behind the celebration of *Navratri* goes that there was a battle between the powerful demon called *Mahishasura* and goddess *Durga*. Lord *Brahma* had blessed *Mahishasura* with immortality with the condition that he can be conquered and killed only by a woman. He became overconfident about his powers and immortality and attacked heaven, hell and earth with the pursuit of conquering them all. No god could stand in front of him as he was immortal. Thereby to save heaven, earth and hell goddess *Durga* was created by Lord *Vishnu* to defeat the demon. The entire powers of Lord *Shiva* and Lord *Brahma* were put together in goddess *Durga* to defeat the demon. Goddess *Durga* is believed to be a power that runs through the universe. Goddess *Durga* fought the battle with the demon for 15 long days. The battle was so terrifying that earth, heaven and hell were on the verge of destruction. Demon *Mahishasura* tried to trick goddess *Durga* by changing forms one after another. But when the demon took the form of a buffalo the goddess *Durga* killed it by using a fork-like weapon called a *trishul* and pierced his chest and killed him instantly.

In *Navratri* the powers of goddess *Durga* in the form of nine *avatars* are worshipped. The first day of *Navratri* is dedicated to worship goddess *Shailputri*. On the second day goddess *Brahmacharini* is worshipped and the third day is dedicated to goddess *Chandraghanta*. On the fourth day goddess *Kushmanda* is worshipped, on the fifth day goddess *Skandamata*, the sixth day goddess *Katyayani*, the seventh day goddess *Kalratri*, and the eighth day goddess *Mahagauri*. On the final ninth day goddess *Siddhidatri* is worshipped. The festival of *Navratri* not only signifies the power of a woman but signifies the victory of good over evil. The majority of people in the Hindu community observe a fast for 9 days in *Navratri*. On the last day they worship small girls in the form of *Kanya Pujan* symbolizing nine *Devis* and give them an invitation to visit their home. They give them gifts, *prasadam*, etc. and perform *pooja* and break their fast.

Not only are there religious sentiments associated with *Navratri* but *Navratri* is taken to be a divine time period to uplift oneself spiritually too. The rituals associated with the festival are in many ways related to detoxifying the mind as well as the body. It is basically dedicated to the feminine aspects of nature in God which is worshipped as divine. The three dimensions of feminine in *Hindutava* are *Durga*, *Lakshmi* and *Saraswati*. The total of nine nights signifies the three dimensions: the first 3 days of *Navratri* are dedicated to goddess *Durga*, the next 3 days are dedicated to goddess *Lakshmi* and the last 3 days are dedicated to goddess *Saraswati*. The *Puranas* (Holy Scriptures in Hinduism) describe the significance and the legend behind *Durga puja* and *Navratri*. Chapters 81–93 of the *Markandeya Purana* (one of the holy scriptures written by Saint *Markandeya*) narrate the entire story of the battle between goddess *Durga* and *Mahishasura* (a demon) and it is referred to as *Devi mahatmya*. The three gods *Brahma*, *Vishnu* and *Shiva* are called the *trimurti* (the trio) and they brought their whole energies together to create an epitome of the power in the form of *Nav Durga* (the nine *avatars* of goddess *Durga*) to win this battle.

Navratri is celebrated twice a year. The first *Navratri* of the year *Chaitra Navratri* (a season as per Hindu mythology), in March–April, describes the victory of Lord *Rama* over *Ravana* with the help of goddess *Durga*, again celebrating womanhood and the victory of good over evil. The second *Navratri* festival is celebrated in the autumn.

Teej

Another prominent festival devoted to a woman (i.e. where the role of a woman is highlighted) is called *Teej*. The festival of *Teej* is celebrated not only in India but also in Nepal. It is dedicated to one of the supreme Indian goddesses, *Parvati*. The festival features the special rich cultural heritage of India and unique folklore, folk music and dance. This festival is mainly celebrated in the months from July to December. *Teej* is celebrated in three different parts: (i) *Haryali Teej*; (ii) *Kajare Teej*; and (iii) *Hartalika Teej*. The legend behind the celebration of the festival is that it is celebrated to mark the marriage of goddess *Parvati* to Lord *Shiva*. The marriage ceremony is regarded to be a significant event in a girl's life in India and that is why the importance of this festival increases in the life of a married woman. Although some critics believe that this is somewhat patriarchal in nature, it has got powerful feminine depiction too. Those who believe in the customs and traditions of married life have many auspicious beliefs associated with this festival. It is basically to mark and celebrate marital devotion and therefore it is considered to be significant as well as auspicious in the life of a woman, which is the axis of life. Without women the world cannot survive and in one way or another after marriage she helps nature to go on.

Basant Panchami

Another festival which depicts devotion to a woman in the form of a goddess is known as *Basant Panchami*. This festival is particularly observed by Hindus in North India and some parts of Nepal. Certain legends of this festival are also associated with the Sikh community. *Basant Panchami* typically marks the beginning of preparations for the spring season. It also signifies starting the preparation for another festival *Holi*, which is generally observed 40 days after *Basant Panchami* where again the role of women is prominent. Hindus dedicate the festival of *Basant Panchami* to the goddess *Saraswati*, the goddess of knowledge, language, music and art. She is the epitome of creative energy and power in the form of art and skill. Goddess *Saraswati* is again the important pillar of the *tridevi* (the trio) (i.e. *Saraswati*, *Lakshmi* and *Parvati*). They are equal forms of the *trimurti: Brahma*, *Vishnu* and *Shiva*, which help to create, maintain and regenerate the universe. As already mentioned, goddess *Saraswati* is believed to grant knowledge, music, spoken wisdom and learning to her devotees. The mention of goddess *Saraswati* can be found in *Rig Veda* (a holy *Hindu* scripture) since the *Vedic* period. She has been regarded as a significant goddess since the ancient period and right up to modern times. Goddess *Saraswati* is depicted as having four arms, one holding a book, one a rosary, another a water pot and one a musical instrument known as *Veena*. All of the items signify symbolically knowledge, music, spoken wisdom and learning. *Vasant Panchami* which is the fifth day of the spring is thereby famously known as *Saraswati Puja* or *Saraswati Jayanti* in honour of goddess *Saraswati*. Many believe that due to the goddess young children can learn to write the letters of the alphabet as she is the goddess of wisdom and learning. Many believers in the ancient Indian religion Jainism also believe in the goddesses *Lakshmi* and *Saraswati*.

Raksha Bandhan

A festival that is dedicated to the bond of a female as a sister is known as *Raksha Bandhan*. Again it is a ceremony that is celebrated by the Hindu community. This is the day where sisters of all ages tie a thread called *rakhi* around the wrists of their brothers as a symbol of protection and they receive a gift in return. This festival generally falls in the month of August, which is the month of *Shravan* in the Hindu lunar calendar. The term *Raksha Bandhan* in Sanskrit literally means the bond of protection, obligation and care which is celebrated in terms of a ritual. It also signifies the importance of a sister in the life of a brother. Without the existence of a female the celebration of this festival again is unimaginable.

Bhaiya Dooj

One more festival in the Hindu culture which celebrates the bond of a brother and sister, again focusing on the importance of a woman as a sister, is *Bhaiya Dooj*. This festival is celebrated in Maharashtra as well as Bengal and is generally observed on the second day of the *Deepavali*. Similar to *Raksha Bandhan* the bond between brother and sister is celebrated. On this day brothers give gifts to their sisters. The whole of Northern India celebrates this festival with great pomp and show. Sisters of all ages put *Kumkum* or *tilika* on their brothers' foreheads. It is also widely celebrated by the Maithili community in Bihar. In Nepal the same festival is celebrated by certain communities. It is also known as *Bhav beej* among Marathi-, Gujarati- and Konkani-speaking communities. There is a popular legend behind the celebration of this festival in Hindu mythology which says that after conquering the evil demon *Narakasura*, Lord *Krishna* visited his sister *Subhadra*, and she gave him a warm welcome with sweets and flowers. She with great affection and love applied *tilika* on *Krishna*'s forehead. Many believe that this was the origin of this festival. Critics also regard this festival as a celebration of patriarchy as the male in the form of a brother is again prominent. But the festival can be viewed from the other angle in the sense that without a sister the ceremonies cannot be performed. The role of a sister is equally as important as that of a brother in the celebration of this festival. The sister's sincerest prayers for the long and happy life of her brother are greeted with respect and love when her brother gives her gifts and cash.

The Modern Qualitative Framework

In the field of social perception, how people create impressions about other people as individual identities are studied. In fact in social perception various social cues are identified in order to make judgments about the roles individuals play in society or the rules governing society. Various situations around us and the prevailing circumstances make us realize and understand about the perceptions of people around us.

One of the important indicators of social behaviour and social patterns these days is social media platforms. What people think, what their perceptions are, and their behaviour towards certain issues or aspects can be judged from the things they post or share on social media platforms, such as Twitter. To analyse what people tweet (i.e. post on Twitter) about women, analysis of Twitter posts (tweets) was performed by using the qualitative data analysis software NVivo. Three keywords were searched for on Twitter to judge the role of women in society.

The first keyword searched for was 'women'. The software captured 17,989 tweets in this search performed on 9 November 2021. The word cloud shown in Fig. 13.1 appeared with the extracted tweets in NVivo. The word cloud is a visual representation of text data. It shows the frequency of the words in the text data. The more common the word within the text being analysed, the larger the word appears in the image generated in the form of the word cloud. Word clouds are extensively used in qualitative research as a tool to identify the focus of the written material. Figure 13.1 shows that in the cluster of words produced from the search conducted on Twitter the main keyword 'women' was the largest, along with words such as: religion, people, India, love, Hindu, prophet, festivals, Muslims. Other than 'women' the most prominent word which is highlighted is 'religion' so we can perceive that to a great extent women are perceived in relation to or with religion. The search word also highlights the concurrence of the word 'festivals' and 'Hindus'.

The second keyword search on Twitter was for 'women and festivals'. The software extracted 1630 tweets and Fig. 13.2 shows the word cloud that was created. The prominent word along with the words 'women' and 'religion' was surprisingly 'festivals'. The search supports the fact that festivals, women and religion are intertwined with each other. In addition to these three words, other prominent words were

Fig. 13.1. Word cloud 1 produced from analysis of Twitter posts featuring the keyword 'women'. From output using NVivo software.

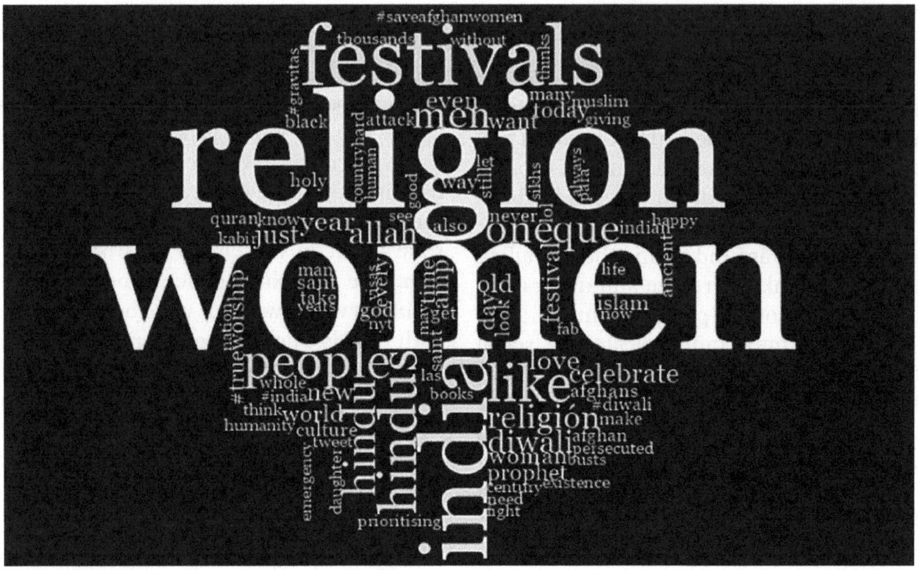

Fig. 13.2. Word cloud 2 produced from analysis of Twitter posts featuring the keyword term 'women and festivals'. From output using NVivo software.

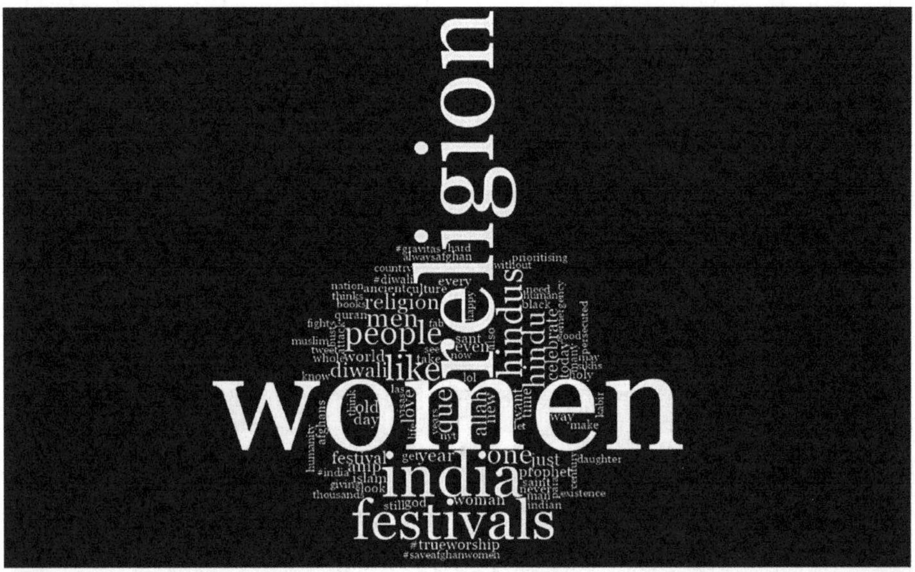

Fig. 13.3. Word cloud 3 produced from analysis of Twitter posts featuring the keyword term 'festivals of India'. From output using NVivo software.

'India', 'people', 'love' and 'Hindus' signifying the linkages of women with festivals in India.

The third keyword search term was 'festivals of India'. This resulted in strong supporting evidence of the importance of women in festivals of India (see Fig. 13.3 showing the large size of the word 'women'). In addition to the keywords 'festivals' and 'India' being among the highest occurring words in the cloud chart, the next most prominent word came out to be 'religion'. We have inferred from the discussions above that up to this time in India religion is directly associated with celebration of festivals, which in turn determines the presence or dominance or signifies the status of women therein. And in this analysis of Twitter posts depicted in word cloud 3 (Fig. 13.3) there is strong evidence of the relationships of women with festivals and religion, indicating their existence in Hindu cultures specifically in India.

Conclusion

It is simply impossible to think of Indian festivals without women. In fact the real colour of most Indian festivals is actually the women. Some critics, though, are of the view that festivals are actually celebrated to exhibit patriarchy. However, as the coin has two faces, in the same way, the author contends that the basic fundamental principle of celebrating life in Indian tradition is actually to give centrality to women. Through festivals, rituals and various religious ceremonies the fact becomes evident that women are inseparable from festivals, the basic essence of these celebrations totally, purely and incredibly relies on the women and their existence is the real basis of these festivals.

References

Hinduism Outlook (n.d.) Ashtalakshmi Stotram Lyrics and Meaning – English. Available at: https://www.hinduismoutlook.com/ashtalakshmi-stotram-lyrics-meaning/ (accessed 25 November 2021).
Mair, J. (ed.) (2018) *The Routledge Handbook of Festivals*. Routledge, Abingdon, UK.
QSR International (n.d.) NVivo, version 1.5.1 (940). Available at: https://www.qsrinternational.com/nvivo-qualitative-data-analysis-software/home (accessed 20 August 2021).
Raman, S.A. (2009) *Women in India: a Social and Cultural History*. ABC-CLIO, Santa Barbara, California.
Sharma, U. (2008) *Festivals in Indian Society*. Mittal Publications, New Delhi.
Wadley, S.S. (1977) Women and the Hindu tradition. *Signs: Journal of Women in Culture and Society* 3(1), 113–125. DOI: 10.1086/493444.

14 An Evaluation of Event Tourism in Terms of Destination Brand Value

Özgür Yayla[1]* and Yiğit Güven[2]
[1]University of Akdeniz, Antalya, Turkey; [2]Ankara Hacı Bayram Veli University, Ankara, Turkey

Introduction

The tourism sector has become very important for many countries. Tourist destinations strive to differentiate themselves from other destinations and provide competitive advantages by focusing on alternative tourism types aside from sea, sand and sun tourism. Geo-economic assets, historical buildings and tourism investments at the destination create the originality of a destination and they are considered among the reasons that motivate tourists to travel (Yayla, 2019). In addition to these, one of the tourism types which has been considered as important for destination branding in recent years is event tourism (Jago *et al.*, 2010). What is more, there is clear evidence in the literature that tourists travelling to attend an event spend more money at the destination, stay longer and travel in groups (Tang and Turco, 2001). In the light of these findings, it can be stated that tourists travelling for events have become a lucrative target market. Considering the economic benefits that all destinations want to achieve, event tourism can be accepted as a very important field in the branding of destinations.

With the increasing importance of event tourism, the number of different events organized in destinations has increased. Although there are studies in the literature in terms of product diversification of event tourism (McKercher, 2016) or the nature of events (Viol *et al.*, 2018), studies on destination branding are few but increasing. This chapter aims to determine the effect of event tourism on the destination brand value using analysis of secondary data. The concepts of event tourism and destination brand are examined comprehensively from studies conducted in the international literature.

The Conceptualization of the Brand Concept

The strengthening of the competitive environment in the globalizing world has made the importance of differentiation very noticeable (Kotler *et al.*, 2007). Branding is one of the most important elements of differentiation, particularly within sectors where there is a highly competitive atmosphere (Aaker, 1991). However, the increase of competition also brings the difficulty of branding and the need for more professional management in terms of branding. In the 21st century, brands are offered to consumers in many and various ways. In this case, it has become necessary to examine the concept of brand academically. Branding, a concept that

*Corresponding author: ozguryayla@akdeniz.edu.tr

DOI: 10.1079/9781789248685.0014

was first discussed in the marketing literature, started to be studied in the 1940s (Guest, 1942). Many authors have made studies in the field of branding and associated the concept of brand with other fields (Aaker, 1991; Kotler, 1994; Keller, 2003; Kim and Kim, 2005).

According to the definition made by the American Marketing Association a 'brand is the name, symbol or sign that distinguishes one product from another' (Kim and Kim, 2005). In another definition, a brand consists of a distinctive name, symbol, logo or sign that aims to identify the goods or services of a seller or a group of sellers and to distinguish these goods from their competitors (Aaker, 1991; Kotler, 1994). What's more, more brands mean more options in the globalizing world. On the condition that some brands are excluded, competition between hundreds of new brands, filling the market, cannot reach serious dimensions unless a considerable brand value can be established (Chang and Liu, 2009). In this context, in order to achieve success, brands should be multidimensional; in other words, their total value should be stronger than that of other brands (Aaker, 1991).

These numerous definitions imply that effective marketing programmes of brand value have an increasing value in the minds of the consumers. This fact increases the trust by encouraging consumers to be loyal and pay a high price for a specific brand (Aaker, 1991). Aaker (1991), considering brand value as a multidimensional concept, stated it as 'A set of brand assets (brand loyalty, brand awareness, brand quality and brand image) attached to the name of a brand or symbol added to or subtracted from the value provided to consumers'. This definition states that brand value adds value to the product or service, increases customer satisfaction and benefits the company.

As for tourism literature, brand value applications have been reported in a few studies (Zenker *et al.*, 2017; Mansilla *et al.*, 2019; Zavattaro and Fay, 2019) and divided into three perspectives: (i) customer focused; (ii) financially focused; and (iii) a combination of the two. However, in general, brand value studies in tourism have focused on Aaker's brand value model (loyalty, awareness, quality and image) (Kim and Kim, 2005). When the model of brand value in Aaker (1991) is analysed in detail, primarily the concept of brand awareness comes up.

Brand Awareness

Brand awareness represents the strength of the brand's presence in the mind of the target mass throughout a certain process (Aaker, 1996). The awareness of a brand is highly regarded by destination management organizations in touristic destinations where competition is intensely experienced. In this context, in order to gain competitive advantage, destination managers aim to raise awareness of the destination by creating a unique brand (Jago *et al.*, 2003).

In the literature, a lot of research has been done in a wide range of areas concerning brand awareness (Oh, 2000). The studies that have been conducted could be exemplified as the effect of brand awareness on: (i) tourism and hotel management (Kotler, 1994; Keller, 2003); (ii) consumer purchasing decisions (Motameni and Shahrokhi, 1998); and (iii) hotel performance (Kim and Kim, 2005). A great many other similar studies have been conducted on the field of tourism, as well (Belonax and Javalgi, 1989; Kwun and Oh, 2004).

Brand Image

Brand image expresses the brand perception in the minds of consumers (Dobni and Zinkhan, 1990; Keller, 2003). When examined in terms of tourism, destination brand image is mainly accepted as the basic dimension of brand value. Blain *et al.* (2005) stated that the destination brand image should be included in the definition of destination brand value. The creation of a brand image constitutes an important component of the destination branding model (Cai, 2002).

When evaluated in terms of travel intentions, brand image among foreign tourists is regarded as primary and valuable data that reflects important indicators of customer segments and market potential and emphasizes the functional, social and emotional brand image in tourism (Camarero *et al.*, 2010). Accordingly, in order to turn these tourists into loyal visitors, organizations aim to create a unique and effective image and identity that will encourage their travel intentions (Kotler, 1994).

A number of different studies have been carried out in order to define and express brand image (Kapferer, 1997; Grace and O'Cass, 2005). In the studies, it has been stated that destination brand image is limited to the social image and self-image of brand personality (Patterson, 1999; Hosany *et al.*, 2006). In some other studies, brand image has also been considered as an element of brand personality (Belk, 1988; de Chernatony and Dall'Olmo Riley, 1998).

Brand Quality

Brand quality is one of the basic dimensions of brand value (Aaker, 1996; Keller, 2003). Brand quality has been used interchangeably with quality perceived by customers (Zeithaml, 1988). In a customer-based brand value model, Keller (2003) identified seven dimensions of product quality. These dimensions were stated as: (i) performance; (ii) features; (iii) conformance quality; (iv) reliability; (v) durability; (vi) ease of service; and (vii) style. Lewis and Chambers (1989) defined perceived quality as a consumer decision resulting from comparisons made by the consumer between expectations and perception of service performance. Boo *et al.* (2009) considered the performance among the seven quality dimensions stated by Keller (2003) as the brand quality. In addition to this, Lassar *et al.* (1995) accepted the brand performance as one of the fundamental dimensions of brand quality. When the destination is examined in terms of brand quality, products such as environment and service infrastructure should be taken into consideration in measuring the destination brand performance (Buhalis, 2000).

Target Brand Loyalty

Brand loyalty is at the centre of brand value (Aaker, 1991; Keller, 2003). Lassar *et al.* (1995) defined brand value as the trust of consumers in a specific brand more than other brands. This leads to loyal consumers and increases their desire to pay for the brand (Murphy *et al.*, 2000). Creating customer loyalty requires a good brand management process. Keller (2003) stated that the main source of customer-based brand value is brand loyalty. In another definition, brand loyalty was defined as a customer's loyalty to the brand (Aaker, 1991; Keller, 2003). Back and Parks (2003) expressed that brand loyalty is seen as a result of multidimensional cognitive attitudes towards a particular brand. When brand loyalty is evaluated from the perspective of tourism, it is very useful for a travel destination (Yoon and Uysal, 2005). Tourists with brand loyalty are more likely to visit that destination again and also they can recommend visiting that destination to those around them (e.g. family, friends, work colleagues) (Baloğlu, 2001). As loyalty is seen as an important research area in tourism (Flavian *et al.*, 2001), studies examining destination brand loyalty within the concept of destination brand value have become quite common.

Konecnik and Gartner (2007) drew attention to the importance of brand loyalty for Slovenia in the brand value model. What is more, Kim and Kim (2005) stated that brand loyalty from customers' perspectives affected performance of tourism businesses. Another study that showed a positive relationship between loyalty and travel intentions is the study carried out by Baloğlu (2001) on consumer behaviour in tourism. Yoon and Uysal (2005) also stated that tourists may feel loyal to a particular brand or destination and may intend to re-buy a branded product or revisit a destination. The future travel intentions of tourists are the result of their perceptions concerning their previous travels (Baloğlu, 2001).

Destination Branding in Tourism

With the globalization of the world, the strengthening of the competitive market causes consumers to encounter many brands. In addition, the marketing practices that brands make in order to differentiate themselves in the competitive environment also target consumers intensively. Therefore, the issue of evaluating the concept of branding has become important for individual sectors, such as the tourism sector.

The concept of branding has been extensively defined in terms of concrete products and intangible service products in the marketing literature (Blain *et al.*, 2005). Branding is also included in the tourism product category; it is possible to see branding in all tourism activities (Cai, 2002; Gnoth *et al.*, 2007). Because tourism consists of tangible and intangible features it can be turned into a brand (Ekinci and Hosany, 2006). In this context, a tourism destination is also a product (Hosany *et al.*, 2007). We have stated in previous sections that a brand is a powerful differentiation tool. Various researchers have stated that differentiation is an important competitive marketing strategy (Kapferer, 1997). The progress of the brand concept from concrete products to the service sector such as tourism has been used as an important differentiation argument for destination management (Buhalis, 2000).

The first research studies on destination branding started in the 1990s. For example, Dosen *et al.* (1998) conducted a study on the branding of Croatia as a destination. In addition to this, there are many studies on destination branding right up until the present time (Ritchie and Ritchie, 1998; d'Hauteserre, 2001). Some studies have been conducted on particular aspects of destination branding such as destination image (Gallarza *et al.*, 2002), destination positioning (Chacko, 1997) and target slogans (Richardson and Cohen, 1993). The studies on how to measure the elements that make up destination branding and what the destination branding components are, are few but increasing (Klenosky and Gitelson, 1997).

Destination branding has become a popular application in the field of tourism and preferred by destination management organizations. Besides, it has become more evident day by day that differentiation and branding should be supported by different activities (Aaker, 1991). In this context, event tourism in the context of tourism organizations is among the preferred activities in order to create this differentiation and to contribute to brand value. This is because the events organized in destinations are important in terms of destination branding.

The Role of Events in Destination Brand Value

As in every sector, the concept of branding is an area that needs to be managed seriously in the tourism sector. In the last 10 years, it has become more evident that destination marketing organizations need to make serious strategic plans in terms of branding. This is because studies indicate that tourists tend to prefer places (i.e. destinations) with high brand power (Boo *et al.*, 2009). This situation shows that branding is as important as other marketing activities. In addition, the branding of the destination provides serious benefits to the consumers in making decisions easier and eliminating various risks (Kotler, 1994). Further, with the rapid advancement of technology, information has become more accessible now, enabling all tourism organizations that care about this situation to carry out their brand studies in a more professional way and also technology is used in how many brand destinations are presented to consumers (Getz, 1998). Many destinations have become brands due to the economic gains that an event held at the destination has brought, and as a result, the decision-making processes of consumers have become more and more difficult because there is so much choice in events/destinations.

Event tourism has entered the tourism literature as a candidate for diversification and branding strategies of destinations. Events can be defined as planned spatial and temporal actions (Persson, 2002). It is possible to say that each event is unique because it differs from any other in terms of environment, participants, management organization and design (Westerbeek *et al.*, 2002). Since the events are never the same, it is necessary to be in the area where the event is organized in order to understand the event (Getz, 2004). However, with the recent development of technology, 'virtual events' are organized using social media and various promotional programmes, especially for branding activities, and different experiences are offered to consumers (Getz *et al.*, 2007), and it is different from traditional event experiences.

All planned events are designed with a purpose, especially in the field of destination and tourism; events are too important to be managed amateurishly. In this context, touristic activities are managed by professionals (Tassiopoulos, 2000; Shone and Parry, 2001). Event management is a field of applied work and a profession focused on the design, production and management of planned events covering many areas such as festivals and other celebrations, entertainment, recreation, political, science, sports and arts (Van der Wagen, 2004). Event tourism is of great importance in terms of regional, national and international branding and the destination image-creation process.

This situation has highlighted the fact that event tourism should be studied as an academic field and that this subject should be organized by educated people. In 1987, the New Zealand Tourist Promotion Department contributed to the emergence of the concept of event tourism by using the expression 'Event tourism has a rapidly growing importance in the spread of international tourism' (Getz and Page, 2016, p. 615). Getz (1989) emphasized the importance of event tourism by drawing a framework for event tourism in an article on tourism management.

The importance of events in destination branding is increasing in the field of tourism. This is because, especially together with regional activities, events in metropolises are considered among the important differentiation elements that help to focus on the region or more local destination rather than the nation as a whole (Crockett and Wood, 1999). With growing interest in destination branding, researchers have recently begun to pay more attention to the impact of festivals and other events on destination branding. A number of studies in the field of festivals and other events, state that festivals are used as a destination branding tool (Chalip and Costa, 2005). In recent years, studies on the contribution of festivals and other events to destination branding have become increasingly common in the literature (Getz, 2008; Stokes, 2008; Mariani and Giorgio, 2017; Kim *et al.*, 2018; Laing, 2018; Mainolfi and Marino, 2020). These activities can be grouped as sports, music, dance, literature or commemorative festivals and food and beverage festivals, which are generally included in event tourism.

Festivals organized in destinations, especially regional festivals, can offer unique contents that reflect the character of the people of the region and facilitate their differentiation, in other words, branding (Wood and Long, 2009). For example, food and beverage festivals, which are one of the common regional events, appear to be easier to brand in the minds of visitors, as they include dishes that reflect the local population (Kalkstein-Silkes *et al.*, 2008). While event tourism adds value to the existing products within the destinations, it also adds value to the branding of the destination and the country (Getz and Brown, 2006).

McCartney (2005) and Jago *et al.* (2003) emphasized that in order for events to be beneficial in terms of destination branding, the activity or event needs to be held at a particular destination and be associated with it for some period of time. They stated that the events contribute to the brand value of the destination in terms of longevity, self-renewal and keeping up with innovations. In addition, Brown *et al.*, in their study in 2002, obtained results that support the literature by saying that festivals organized for more than 10 years leave more permanent effects on destination branding. Also Brown *et al.* (2002) stated that one-off mega-events such as the Olympics contribute positively to destination branding on a national scale. However, this situation is not valid for the activities/events that are carried out on a smaller scale such as in a particular local region or local destination – these only contribute to local destination branding. It has also been stated that regular events would be more beneficial in order to contribute to local branding, to make local products popular and to ensure sustainable branding (Boo *et al.*, 2009).

Conclusion and Suggestions

In this research, the concepts of brand and branding, brand value and dimensions, destination branding in tourism, and the effects of event tourism on destination brand value, have been examined in the light of the secondary data. The concept of branding, with the developing technology and opportunities, has now become a necessity not only at the level of countries but at the level of institutions and organizations, as well. Differentiation and being a brand play a vital role for businesses and destinations, considering the efforts of many brands to compete in a competitive environment and to determine the preferences of consumers.

It is an undeniable fact that destinations are in competition with each other just like companies and organizations. However, in the 21st century, touristic destinations have to operate in various fields in order to better reflect their cultural identity and create their own branding. The field of tourism is a field of study that has a national impact and even has the capacity to differentiate in the field of international competition. The promotion, management and organization of both tangible and intangible assets of tourism contribute to the image of the country and enable regional development.

In this context, planned and professional activities to be created within tourism will significantly contribute to the brand value both regionally and nationally. Events, festivals and other organizations in tourism have a very important position in terms of destination branding. Activities carried out at a destination within the scope of event tourism contribute to the branding of the region and even cause the surrounding locations to benefit from this branding.

It is important for destination management organizations to work with professional event managers in events (festivals, etc.) to strengthen the destination branding and to organize events that reflect the identity of the destination. As stated above, the effect of the events on the destination brand value depends on the harmony of the dynamics between the event and the destination. In addition, it can be said that one-off regional events will not contribute to the brand value of a destination; instead events should be organized to take place, for example, at a certain time of year, every year for a long time period in order for the development of brand value and the formation of a sustainable brand.

References

Aaker, D.A. (1991) *Managing Brand Equity*. Free Press, New York.

Aaker, D.A. (1996) Measuring brand equity across products and markets. *California Management Review* 38(3), 102–120.

Back, K.-J. and Parks, S.C. (2003) A brand loyalty model involving cognitive, affective, and conative brand loyalty and customer satisfaction. *Journal of Hospitality & Tourism Research* 27(4), 419–435.

Baloğlu, S. (2001) An investigation of a loyalty typology and the multidestination loyalty of international travelers. *Tourism Analysis* 6(1), 41–52. DOI: 10.3727/108354201108749737.

Belk, R.W. (1988) Possessions and the extended self. *Journal of Consumer Research* 15(2), 139–168.

Belonax, J.J. and Javalgi, R.G. (1989) The influence of involvement and product class quality on consumer choice sets. *Journal of the Academy of Marketing Science* 17(3), 209–216.

Blain, C., Levy, S.E. and Ritchie, J.R.B. (2005) Destination branding: insights and practices from destination management organizations. *Journal of Travel Research* 43(4), 328–338. DOI: 10.1177/0047287505274646.

Boo, S., Busser, J. and Baloglu, S. (2009) A model of customer-based brand equity and its application to multiple destinations. *Tourism Management* 30(2), 219–231.

Brown, G., Chalip, L., Jago, L. and Mules, T. (2002) The Sydney Olympics and brand Australia. In: Morgan, N., Pritchard, A. and Pride, R. (eds) *Destination Branding: Creating the Unique Destination Proposition*. Butterworth-Heinemann, Oxford, UK, pp. 163–185.

Buhalis, D. (2000) Marketing the competitive destination of the future. *Tourism Management* 21(1), 97–116.

Cai, L. (2002) Cooperative branding for rural destinations. *Annals of Tourism Research* 29(3), 720–742.

Camarero, C., Garrido, J.M. and Vicente, E. (2010) Components of art exhibition brand equity for internal and external visitors. *Tourism Management* 31(4), 495–504. DOI: 10.1016/j.tourman.2009.05.011.

Chacko, H.E. (1997) Positioning a tourism destination to gain a competitive edge. *Asia Pacific Journal of Tourism Research* 1(2), 69–75.

Chalip, L. and Costa, C. (2005) Sport event tourism and the destination brand: towards a general theory. *Sport in Society* 8(2), 218–237. DOI: 10.1080/17430430500108579.

Chang, H.H. and Liu, Y.M. (2009) The impact of brand equity on brand preference and purchase intentions in the service industries. *The Service Industries Journal* 29(12), 1687–1706.

Crockett, S.R. and Wood, L.J. (1999) Brand Western Australia: a totally integrated approach to destination branding. *Journal of Vacation Marketing* 5(3), 276–289.
de Chernatony, L. and Dall'Olmo Riley, F. (1998) Defining a brand: beyond the literature with experts' interpretations. *Journal of Marketing Management* 14(5), 417–443.
Dobni, D. and Zinkhan, G.M. (1990) In search of brand image: a foundation analysis. In: Goldberg, M.E., Gorn, G. and Pollay, R.W. (eds) *Advances in Consumer Research*, Vol. 17. Association for Consumer Research, Provo, Utah, pp. 110–119.
Dosen, D.O., Vranesevic, T. and Prebezac, D. (1998) The importance of branding in the development of marketing strategy of Croatia as tourist destination. *Acta Turistica* 10(2), 93–182.
d'Hauteserre, A.M. (2001) Destination branding in a hostile environment. *Journal of Travel Research* 39(3), 200–207. DOI: 10.1177/004728750103900308.
Ekinci, Y. and Hosany, S. (2006) Destination personality: an application of brand personality to tourism destinations. *Journal of Travel Research* 45(2), 127–139.
Flavian, C., Martinez, E. and Polo, Y. (2001) Loyalty to grocery stores in the Spanish market of the 1990s. *Journal of Retailing and Consumer Services* 8(2), 85–93. DOI: 10.1016/S0969-6989(99)00028-4.
Gallarza, M.G., Saura, I.G. and Garcia, H.C. (2002) Destination image – towards a conceptual framework. *Annals of Tourism Research* 29(1), 56–78.
Getz, D. (1989) Special events: defining the product. *Tourism Management* 10(2), 125–137.
Getz, D. (1998) Event tourism and the authenticity dilemma. In: Theobald, W. (ed.) *Global Tourism*, 2nd edn. Butterworth-Heinemann, Oxford, UK, pp. 409–427.
Getz, D. (2004) Geographic perspectives on event tourism. In: Lew, A., Hall, M. and Williams, A. (eds) *A Companion to Tourism Geographic Perspectives on Event Tourism*. Blackwell Publishing, Oxford, UK, pp. 410–422.
Getz, D. (2008) Event tourism: definition, evolution, and research. *Tourism Management* 29(3), 403–428. DOI: 10.1016/j.tourman.2007.07.017.
Getz, D. and Brown, G. (2006) Critical success factors for wine tourism regions: a demand analysis. *Tourism Management* 27(1), 146–158.
Getz, D. and Page, S.J. (2016) Progress and prospects for event tourism research. *Tourism Management* 52, 593–631.
Getz, D., Andersson, T. and Larson, M. (2007) Festival stakeholder roles: concepts and case studies. *Event Management* 10(2/3), 103–122.
Gnoth, J., Baloglu, S., Ekinci, Y. and Sirakaya-Turk, E. (2007) Introduction: building destination brands. *Tourism Analysis* 12(5), 339–343.
Grace, D. and O'Cass, A. (2005) Service branding: consumer verdicts on service brands. *Journal of Retailing and Consumer Services* 12(2), 125–139.
Guest, L.P. (1942) The genesis of brand awareness. *Journal of Applied Psychology* 26(6), 800–808. DOI: 10.1037/h0055390.
Hosany, S., Ekinci, Y. and Uysal, M. (2006) Destination image and destination personality: an application of branding theories to tourism places. *Journal of Business Research* 59(5), 638–642.
Hosany, S., Ekinci, Y. and Uysal, M. (2007) Destination image and destination personality. *International Journal of Culture, Tourism and Hospitality Research* 1(1), 62–81.
Jago, L., Chalip, L., Brown, G., Mules, T. and Ali, S. (2003) Building events into destination branding: insights from experts. *Event Management* 8(1), 3–14.
Jago, L., Dwyer, L., Lipman, G., van Lill, D. and Vorster, S. (2010) Optimising the potential of mega-events: an overview. *International Journal of Event and Festival Management* 1(3), 220–237.
Kalkstein-Silkes, C., Cai, L.A. and Lehto, X.Y. (2008) Conceptualizing festival-based culinary tourism in rural destinations. In: Hall, C.M. and Sharples, L. (eds) *Food and Wine Festivals and Events Around the World*. Butterworth-Heinemann, Oxford, UK, pp. 65–77.
Kapferer, J.N. (1997) *Strategic Brand Management: Creating and Sustaining Brand Equity Long Term*, 2nd edn. Kogan Page, London.
Keller, K.L. (2003) *Strategic Brand Management*. Prentice-Hall, Upper Saddle River, New Jersey.
Kim, H.B. and Kim, W.G. (2005) The relationship between brand equity and firms' performance in luxury hotels and chain restaurants. *Tourism Management* 26(4), 549–560.
Kim, S., Choe, J.Y. and Petrick, J.F. (2018) The effect of celebrity on brand awareness, perceived quality, brand image, brand loyalty, and destination attachment to a literary festival. *Journal of Destination Marketing & Management* 9, 320–329.

Klenosky, D.B. and Gitelson, R.E. (1997) Characteristics of effective tourism promotion slogans. *Annals of Tourism Research* 24(1), 235–238.

Konecnik, M. and Gartner, W.C. (2007) Customer-based brand equity for a destination. *Annals of Tourism Research* 34(2), 400–421.

Kotler, P. (1994) *Marketing Management: Analysis, Planning, Implementing and Control*. Prentice-Hall, Upper Saddle River, New Jersey.

Kotler, P., Brown, L., Adam, S., Burton, S. and Armstrong, G. (2007) *Marketing*, 7th edn. Pearson Education, Frenchs Forest, New South Wales, Australia.

Kwun, J.W. and Oh, H. (2004) Effects of brand, price, and risk on customers' value perceptions and behavioral intentions in the restaurant industry. *Journal of Hospitality & Leisure Marketing* 11(1), 31–49.

Laing, J. (2018) Festival and event tourism research: current and future perspectives. *Tourism Management Perspectives* 25, 165–168.

Lassar, W., Mittal, B. and Sharma, A. (1995) Measuring customer-based brand equity. *Journal of Consumer Marketing* 12(4), 11–19.

Lewis, R.C. and Chambers, R.E. (1989) *Marketing Leadership in Hospitality*. Van Nostrand Reinhold, New York, pp. 313–318.

Mainolfi, G. and Marino, V. (2020) Destination beliefs, event satisfaction and post-visit product receptivity in event marketing. Results from a tourism experience. *Journal of Business Research* 116, 699–710.

Mansilla, Ó., Contrí, G. and Cantallops, A. (2019) The impact of value co-creation on hotel brand equity and customer satisfaction. *Tourism Management* 75, 51–65.

Mariani, M.M. and Giorgio, L. (2017) The "Pink Night" festival revisited: meta-events and the role of destination partnerships in staging event tourism. *Annals of Tourism Research* 62, 89–109.

McCartney, G.J. (2005) Hosting a recurring mega-event: visitor raison d'etre. *Journal of Sport Tourism* 10(2), 113–128.

McKercher, B. (2016) Towards a taxonomy of tourism products. *Tourism Management* 54, 196–208.

Motameni, R. and Shahrokhi, M. (1998) Brand equity valuation: a global perspective. *Journal of Product & Brand Management* 7(4), 275–290.

Murphy, P., Pritchard, M.P. and Brock Smith, J. (2000) The destination product and its impact on traveller perceptions. *Tourism Management* 21(1), 43–52.

Oh, H. (2000) Diners' perceptions of quality, value, and satisfaction. *Cornell Hotel and Restaurant Administration Quarterly* 41(3), 58–66.

Patterson, M. (1999) Re-appraising the concept of brand image. *Journal of Brand Management* 6(6), 409–426.

Persson, C. (2002) The Olympic Games site decision. *Tourism Management* 23(1), 27–36.

Richardson, J. and Cohen, J. (1993) State slogans: the case of the missing USP. *Journal of Travel & Tourism Marketing* 2(2/3), 91–109.

Ritchie, J.R.B. and Ritchie, R.J.B. (1998) The branding of tourism destinations: past achievements and future challenges. In: Keller, P. (ed.) *Proceedings of the 1998 Annual Congress of the International Association of Scientific Experts in Tourism, Destination Marketing: Scopes and Limitations*. International Association of Scientific Experts in Tourism, Marrakech, Morocco, pp. 89–116.

Shone, A. and Parry, B. (2001) *Successful Event Management*. Continuum, London.

Stokes, R. (2008) Tourism strategy making: insights to the events tourism domain. *Tourism Management* 29(2), 252–262.

Tang, Q. and Turco, D.M. (2001) Spending behaviors of event tourists. *Journal of Convention & Exhibition Management* 3(2), 33–40.

Tassiopoulos, D. (ed.) (2000) *Event Management: A Professional and Developmental Approach*. Juta Education, Lansdowne, South Africa.

Van der Wagen, L. (2004) *Event Management for Tourism, Cultural, Business and Sporting Events*, 2nd edn. Pearson Education, Frenchs Forest, New South Wales, Australia.

Viol, M., Todd, L., Theodoraki, E. and Anastasiadou, C. (2018) The role of iconic-historic commemorative events in event tourism: insights from the 20th and 25th anniversaries of the fall of the Berlin Wall. *Tourism Management* 69, 246–262. DOI: 10.1016/j.tourman.2018.06.018.

Westerbeek, H., Turner, P. and Ingerson, L. (2002) Key success factors in bidding for hallmark sporting events. *International Marketing Review* 19(3), 303–322.

Wood, E.H. and Long, P. (2009) Great Yorkshire shows: the role of festivals in regional tourism development. In: Thomas, R. (ed.) *Managing Regional Tourism: a Case Study of Yorkshire, England*. Great Northern Books, Ilkley, UK.

Yayla, Ö. (2019) Factors affecting the tourist purchase behaviour. In: Krystev, V., Efe, R. and Atasoy, E. (eds) *Theory and Practice in Social Sciences*. St. Kliment Ohridski University Press, Sofia, Bulgaria, pp. 580–591.

Yoon, Y. and Uysal, M. (2005) An examination of the effects of motivation and satisfaction on destination loyalty: a structural model. *Tourism Management* 26(1), 45–56. DOI: 10.1016/j.tourman.2003.08.016.

Zavattaro, S. and Fay, D. (2019) Brand USA: a natural quasi-experiment evaluating the success of a national marketing campaign. *Tourism Management* 70(6), 42–48. DOI: 10.1016/j.tourman.2018.07.013.

Zeithaml, V.A. (1988) Consumer perceptions of price, quality, and value: a means-end model and synthesis of evidence. *Journal of Marketing* 52(3), 2–22.

Zenker, S., Braun, E. and Petersen, S. (2017) Branding the destination versus the place: the effects of brand complexity and identification for residents and visitors. *Tourism Management* 58, 15–27.

15 Technology and Crowd Management at Events: a Case Study of Kumbh Festival in India

Pinaz Tiwari* and Nimit Chowdhary
Jamia Millia Islamia, New Delhi, India

Introduction

Religious tourism is an integral part of the tourism industry. Several developing economies offer religious tourism activities on a larger scale (Vučetić, 2021), and religious destinations attract more than 300 million visitors each year worldwide. This conventional form of tourism has been a part of mass tourism for decades and encompasses pilgrimage sites, religious and spiritual tourists, religious events and ritual processions. Religion-based events are one of the dimensions of religious tourism that garners large crowds at the sites. Several religious practitioners and people with faith and hedonistic motivations visit the events to experience and participate in the rituals. Often termed mass gathering events (MGEs), these events are organized in a definite space. However, the wide-scale popularity and religious faith of communities often have negative impacts on these events. The issues of overcrowding, lack of management, and unsustainable use of resources are predominant at religious events, especially in India.

India is famous for religious festivals and sites that attract millions of people worldwide. One such festival is the Kumbh festival. It is also considered one of the world's largest human gatherings and is organized in different Indian cities. In the past, various instances of stampedes, accidents and the transmission of infections have been reported (Yamin, 2019; Aggrawal *et al.*, 2020; Quadri and Padala, 2021). These incidents maximize the chances of public health concerns at the MGEs.

Being the world's second most populated country and people belonging to diverse cultures and religions, crowding issues are inevitable at popular religious sites in India. Moreover, religious tourism and related events are much inclined towards attracting the mass market. As the market is becoming a part of globalization in the post-modernism era (Iliev, 2020), crowd management is necessary for:

- ensuring the safety of the staff and visitors;
- combatting the risks of public health and accidents; and
- improving visitors' quality of experience.

The Kumbh festival faced widespread criticism from India and abroad for being conducted amid the coronavirus disease 2019 (COVID-19) pandemic. Thousands of people did not adhere to COVID-19 protocols, and inadequate crowd-monitoring steps caused an overwhelming situation. Eventually, the COVID-19 positive cases increased drastically and caused the second

*Corresponding author: tiwaripinaz@yahoo.com

DOI: 10.1079/9781789248685.0015

wave in the country. Therefore, monitoring the flow of visitors and managing religious events is a prerequisite to combat the underlying issues.

Considering the revolution of information and communication technology (ICT) in products and services, it can resolve the crowding issues. Indeed, using ICT through intelligent devices and advanced technologies at the event sites would optimize management efficiency (Saha *et al.*, 2020) and re-engineer the site management processes in the longer run. Therefore, using ICT to manage and monitor visitors in the post-pandemic era would be essential to mitigate the chances of infectious diseases and concerns related to crowding.

The contemporary scholarly works on COVID-19 and religious events are inadequate. Considering the abrupt nature of the pandemic, several studies have begun to re-discuss the forgotten pillars of sustainability in tourism (Chang *et al.*, 2020). Furthermore, overcrowding and mass tourism activities have been criticized lately (Goodwin, 2017) in both academia and the media. However, MGEs rarely face criticism even though crowd density accelerates the transmission of diseases (Ebrahim *et al.*, 2020) and mass gatherings work as a playground for the viruses (Elachola *et al.*, 2014). Amid the COVID-19 wave, crowd management at religious sites such as Mecca, Saudi Arabia, during the Hajj pilgrimage, was appreciated for carrying out strategies that ensured the safety of the staff and visitors. On the other hand, the Kumbh (an event on the same scale as the Hajj) was considered a 'super-spreader' incident in the history of the COVID-19 pandemic in the country (Tabish, 2021). Certainly, mass events attract visitors from different parts of the world. Owing to their popularity, people's faith and religious significance, crowding at large-scale events such as the Kumbh cannot be avoided entirely. Thus, crowd management becomes vital for reducing public health risks and visitors' quality of experience.

This study aims to analyse the scope of managing MGEs, especially the Kumbh using ICT. Accordingly, this chapter has two objectives. First, it provides a descriptive analysis of religious events and sustainability-related issues by considering the case of the Kumbh festival. Second, it suggests incorporating ICT to deal with crowd density and health issues that would become a prerequisite for visitors in the post-COVID-19 phase. Given the complexities of the religious tourism sector (Iliev, 2020), this study provides a multidisciplinary approach. It comprehends the dynamics of crowding and using technology to combat the issues at large-scale religious congregations. With the involvement of diverse stakeholders in organizing the Kumbh festival, the scope of the study is broad.

Religious Events and the Case of the Kumbh Festival

Religious tourism implies travelling for religion which includes 'religious duties' experiencing 'religious form' or 'religious practice' or products that prompt art, culture or architecture (Griffin and Raj, 2017). Even though, the idiosyncratic differences between religious travellers, pilgrim travellers and tourists are elusive (Collins-Kreiner, 2020). An extensive number of studies have been conducted to analyse different paradigms related to religious tourism, especially in India (Singh, 2002; Shinde, 2012; Saha *et al.*, 2020). India is considered the birthplace of prominent religions and religious leaders. The number of people participating in religious gatherings is often higher than those participating in political gatherings or other entertainment activities (Gayathri *et al.*, 2017). As one of the conventional forms of human mobility, millions of Indians are still motivated by their religious sentiments to travel to holy sites. Broadly, religious-based tourism includes various activities such as visiting pilgrim sites, attending faith-based events and gatherings, and missionary travels (Tomljenović and Dukić, 2017). The Kumbh event is a faith-based human gathering in India that has been attended by millions of people for decades.

The Kumbh festival has a mythological significance in Hinduism. According to the religion's scriptures, a fight took place between the demons and the Gods to possess the pitcher that contained the elixir of life. As the struggle to obtain the pitcher intensified, a few drops from the pitcher spilt on the Earth. It is said that the drops spilt at four places which currently serve as the venues of the Kumbh festival. These are Ujjain in Madhya Pradesh (Shipra River), Nashik in Maharashtra (Godavari River),

Prayagraj in Uttar Pradesh (Ganges River) and Haridwar in Uttarakhand (Ganges River). The struggle to possess the elixir continued for 12 days, which signifies the cycle of the Kumbh event at four venues in 12 years. However, at Haridwar and Prayagraj, the event takes place twice in 12 years, in a gap of 6 years each. Hence, the event is famously known as Ardh Kumbh (literally meaning half Kumbh) when organized in 6 years, and Purna Kumbh (literally meaning full Kumbh) when organized in 12 years. Due to the significance of the event and representation of cultural integrity, the United Nations Educational, Scientific and Cultural Organization (UNESCO) added the event to the List of Intangible Cultural Heritage of Humanity in 2017 (UNESCO, 2017).

Quadri and Padala (2021) stated that the festival reflects the sociocultural significance of the Hinduism religion. Culturally rich, and spiritually inclined, the Kumbh festival attracts people from diverse backgrounds. Out of the seven ritual practices at the festival, *Snan*, *Yajnas* and *Deep Dan* are performed at specific places, on specific days and times. Moreover, *Snan* is performed on a designated day when millions of people take a dip in the holy river. Due to its auspiciousness (because of the drop of elixir in the river), the ritual is full of religious fervour (Quadri and Padala, 2021). Consequently, large crowds gather to perform these rituals due to restricted space and limited time. Furthermore, people push each other to make way for themselves and their families.

This mega gathering event is organized on a large scale where state government, local official bodies, stakeholders from different domains (such as catering, accommodation and security), volunteers and a few non-governmental organizations come together. Facilities are provided for the pilgrims but these are to meet their basic needs and are not necessarily directed at improving visitors' experiences at the event. In 2013, 11 sub-zones across the site of the festival, each with more than 50 services, were set up; the services included ambulance, sanitation, first aid, potable water and food (Dwivedi and Cariappa, 2015).

Various studies have attempted to study the crowd dynamics concerning the Kumbh festival (Karthika *et al.*, 2018; Verma and Sarangi, 2019; Quadri and Padala, 2021). Discussing the Kumbh event at Ujjain in 2016, Gayathri *et al.* (2017) pointed out the inadequacy of pre-planned crowd control measures and the inconsistency in managing the crowd. The authors highlighted that crowd control steps like barricading and restriction of vehicular movements were implemented only when crowd densities exceeded manageable levels. Also, a few routes were changed without any prior information or appropriate signages that further caused perplexity among visitors.

The Kumbh event that took place in Haridwar amid the COVID-19 pandemic in April 2021 was different from the preceding Kumbh events, especially in terms of safety precautions undertaken by the authorities. Although earlier studies reported the threat of contagious diseases and infections due to communal bathing practices at religious events (Aggrawal *et al.*, 2020; Clara-Rahola, 2020), the government and authorities failed to acknowledge the underlying menace. India witnessed the first COVID-19 wave in October 2020, and the chances of COVID-19 spread could not be ruled out by the event authorities. Therefore, considering the auspiciousness of the festivals, the state government decided to schedule the MahaKumbh event from 14 January to 27 April 2021 (Manral, 2020). Subsequently, dates for *Snans* were also announced, especially the *Shahi Snans* (includes four major *Snans*) to be held in March and April 2021.

Nevertheless, the leading body responsible for managing the event assured that precautionary measures would be taken to mitigate the public health risks concerned with mass gatherings amid the COVID-19 pandemic. Quadri and Padala (2021) also suggested several measures to impede the possibility of a health crisis having roots from the famous MahaKumbh. Some of Quadri and Padala's (2021) proposals are as follows:

- limiting the number of pilgrims – significantly restricting senior citizens and sick people from attending the festival;
- introducing an online registration system and allowing only registered pilgrims at the site;
- collaborating with religious leaders to spread public health messages and awareness;

- employing GPS (Global Positioning System) tracking systems, drones and infographics; and
- ensuring adequate infrastructure for quarantining, healthcare facilities, and installing sanitization stations at the festival.

Despite the detailed list of COVID-19 guidelines prepared by the leading authorities (Bansal, 2021), the reality on the ground seemed to show a contrasting picture. Khare (2021) reported that people did not adhere to the COVID-19 guidelines and the authorities did not follow any stringent checking measures. As a result, people were seen without masks, thousands of devotees took the holy dip in a crowded space, and the infrastructure was falling apart. The Kumbh was ultimately called off on 17 April 2021, due to wide-scale criticism from the general public and medical authorities in India.

Crowding and Social Interference Theory

Crowding dynamics and people's motivation to travel to a place is understood through the social interference theory. Stokols (1972) and Schmidt and Keating (1979) initially discussed the theory to demonstrate the negative psychological assessment of physical spaces. The social interference theory assumes that a traveller's behaviour is motivated by a specific goal (seeking spirituality, engaging in rituals and traditions, taking a holy dip as part of the religious duty). Subsequently, the traveller would negatively evaluate the crowd when his/her motivation to visit the event gets obstructed by the presence of other travellers (Lee and Graefe, 2003). For example, a traveller visited the Kumbh to seek an immersive experience at the riverbank. However, the presence of other pilgrims is confronting his travel motive and thus is leading to higher perceptions of crowding for that traveller. However, it is equally crucial to note that nationalities play an important role in forming negative evaluations of crowding (Sun and Budruk, 2015). For example, as Asians are less vulnerable to crowding (Neuts and Nijkamp, 2012), the visitors at the Kumbh perceive crowding as usual.

Managing the Kumbh Festival using ICT in the Post-pandemic Phase

Religion is a part of society's belief system, and religious rituals are deep-rooted among different communities (Ebrahim *et al.*, 2021). The sustenance of these collective religious rituals and traditions creates a positive emotional, mental and spiritual well-being among people. Any upheaval at these events is observed as unfavourable and hurts the religious beliefs of a community at large. It may further result in geopolitical consequences at the destination. Thus, it is vital to organize large-scale religious events with empathy towards visitors. A well-drafted plan is required with intelligent use of technology and information to ensure proper management.

The Kumbh festival of 2021 was tagged as the super-spreader of coronavirus in the country. Despite surveillance and other preventive measures that had been undertaken by the authorities, crowd management at the site is a challenging phenomenon. This crowding is mainly due to the high collective spiritual quotient among the visitors. However, as the Kumbh festival will be continued in the post-pandemic phase, it is crucial to take preventive measures to monitor the crowd using ICT. This intervention will help mitigate the unpredictable health risks at the event and subsequently lead to sustainability at Kumbh and similar MGEs in India. Some of the suggestions are discussed below.

(a) Incorporating ICT with demarketing strategies

Demarketing is an emerging phenomenon in the tourism sector. Various destination management authorities and national organizations have adopted it to control the negative impacts of tourism at a destination. Crowding is one of the negative impacts, and demarketing is considered a suitable strategy for short-term sustainability. Tiwari *et al.* (2020) suggested implementing demarketing strategies using the '7 Ps' of the marketing mix to ensure destination management and enhance tourists' experience. The strategy is viable and effective in crises like COVID-19. For instance, the Saudi Arabian Government limited the number of pilgrims on the Hajj in 2020 by encouraging online application (place demarketing). It further gave preference to the residents of Saudi Arabia who had no COVID symptoms and those who had not visited the holy site before (product demarketing).

(b) Using ICT to monitor the crowd at sub-zones

Site management at the Kumbh requires precision in controlling different locations within the event site. People's perception of crowding differs as they might enjoy the presence of others at different sub-zones (Mowen *et al.*, 2003). Therefore, sub-zone management is a prerequisite for more extensive management of massive events like Kumbh. Different rituals take place at the Kumbh festival. *Aarti*, *Langar*, *Kirtan* and the *Pravachana* are held regularly, while the *Yajnas*, *Deep Dan* and *Snans* are performed on specific days, at a specific time and place. A large crowd gathers to perform the *Yajnas*, *Deep Dan* and *Snans* as they are time-bound and take place at designated days and places. The chances of a stampede and accidents are more likely to occur during these rituals. Therefore, separate plans are required to manage the crowd during these rituals to: (i) avoid accidents and a stampede; (ii) mitigate the infection rate and other health risks; and (iii) allow visitors to experience spirituality. Even though drones and surveillance cameras monitor the crowd at the event, it is advisable to draft strategies to monitor sub-zones more efficiently.

(c) Foot mapping and crowd surveillance

The Kumbh festival has often been in the limelight for the mishaps and stampede-like situations caused by the large gatherings. The authorities have attempted to combat the problem but this ended in vain, until an 18-year-old teenager, Nilay Kulkarni, invented an electrical mat that counted the footfall of people who step on the mat at a specific time (Pal, 2018). The mat was used in the Nashik Kumbh festival in 2015. This foot-mapping device assisted in monitoring the crowd flow and analysing the risk involved of crowding at a large-scale event. The use of foot-mapping devices in mega-events such as Kumbh is another ICT-based solution to monitor the crowd, and consequently avoid stampedes. Furthermore, the invention made by the teenager is an apt example of sustainability and social entrepreneurship among the youngest generation in the events sector (Bhalla *et al.*, 2021).

In cultural events, it is challenging to avoid crowd experiences, especially when the event is popular. The probability of risking individuals' lives due to accidents and other issues is high (Ahmed and Memish, 2019), particularly during the pandemic. Furthermore, crowd surveillance using advanced technology would help monitor the crowd efficiently and analyse the real-time behaviour of visitors (Khan *et al.*, 2020).

(d) Mobile app registration

App-based registration has become a common phenomenon in India in the recent past. The government aims to build digital infrastructure and make India a leading economy by providing ICT with support from several public- and private-sector organizations. Furthermore, the COVID-19 vaccinations were also based on app registrations.

Since the Kumbh festival does not require pilgrims to register beforehand, it leads to a lack of data on pilgrims visiting the event (Yamin, 2019). Consequently, the officials either over-estimate or under-estimate the flow of pilgrims visiting the Kumbh. Therefore, it is suggested that the official authorities launch mobile applications to register for mega-events like Kumbh. This step will help achieve accuracy in calculating the number of visitors and lead to better monitoring. These apps can also be utilized to offer valuable insights, safety precautions and event timings to visitors in their preferred languages. However, it is equally crucial to consider the segment of people who lack digital literacy to operate a mobile app. Therefore, on-site registration facilities could remain the alternative.

(e) Advanced healthcare system using telemedicine communication

Telemedicine implies healing at a distance and is a natural evolution of the healthcare system in the digital world (American Telemedicine Association, 2020). The idea of telemedicine is accepted in various developed economies as it reduces the burden on hospitals and local clinics. However, in large events like Kumbh, the event management authorities often seem perplexed when it comes to offering healthcare facilities to visitors. In 2021, millions of people attended the Kumbh event, leading to higher chances of spreading the coronavirus infection and other diseases amid the COVID-19 pandemic. Telemedicine as an alternative solution would be helpful for visitors who report minor health issues.

Conclusion

Being one of the MGEs, the Kumbh festival requires advanced solutions. The existing studies related to Kumbh highlighted the crowd dynamics at different cities where the festival is conducted. This chapter has described the significance of the Kumbh in Hindu mythology, crowding issues and how the situation unfolded during the pandemic. The crowding issue, the non-adherence to the COVID-19 protocols by pilgrims and insufficient monitoring of the event eventually meant that the festival became a super-spreader of COVID-19 infection in the country. The authors highlighted that a religious mega-event like Kumbh that attracts millions of people worldwide and holds spiritual significance for thousands of pilgrims could not be averted. Conceivably, large religious congregations have economic, spiritual and geopolitical implications on different stakeholders and the general public. Moreover, although the world was struggling with COVID-19, the state government of Uttarakhand scheduled significant rituals that were part of Kumbh in March and April 2021. For a country known for its religious diversity, managing and monitoring the crowd at the Kumbh festival remains the ultimate solution for the official authorities.

Respective state governments allocate an enormous amount of funds to organizing the Kumbh event. As a result, wide-scale promotions, on-site infrastructure for accommodating visitors, cleaning rivers, food, sanitation facilities and other amenities are taken care of. In addition, advanced technology and communication tools have been widely used at the Kumbh festivals for the last few years. Nevertheless, the COVID-19 pandemic amplified the inadequacies concerning adherence to protocols and crowd management at the MGE. Accordingly, the study in this chapter identified five ICT-based solutions to manage the crowds more efficiently at the festival.

The post-COVID-19 scenario demands radical reforms in the manner events are being organized and managed. Therefore, creating an effective and efficient digital infrastructure would be a prerequisite in managing the crowd at mega-events. The authors identified that incorporating demarketing strategies with ICT, foot mapping, sub-zone management of place using ICT, telemedicine facilities and mobile app registration would assist in better management of Kumbh in the post-COVID-19 phase. These solutions mitigate the unsustainability concerns at religious mega-events like Kumbh and acknowledge the potential of ICT in reducing public health risks.

The probability of COVID-19 spread and subsequent waves cannot be ruled out in India. Only a small percentage of Indians were fully vaccinated when states started to unlock after the second wave. However, despite a massive vaccine drive being implemented by the government, the third wave of COVID-19 driven by the Omicron variant hit the country in late 2021–early 2022. The current situation in February 2022 indicates the necessity for citizens to receive booster shots in a phased manner. Nevertheless, it is assumed that similar religion-based MGEs will again take place in future owing to the deep religious sentiments shared by people in India. Perhaps, the unruly behaviour will resurface at social and religious gatherings. Therefore, it is suggested that the government and the leading bodies responsible for organizing similar events should develop and adopt digital tools for better management of events. Sustained and strict vigilance is imperative so that no religious event gets the title 'super-spreader' of infections in the future.

References

Aggrawal, V., Dikid, T., Jain, S.K., Pandey, A., Khasnobis, P. *et al.* (2020) Disease surveillance during a large religious mass gathering in India: the Prayagraj Kumbh 2019 experience. *International Journal of Infectious Diseases* 101(2020), 167–173. DOI: 10.1016/j.ijid.2020.09.1424.

Ahmed, Q.A. and Memish, Z.A. (2019) From the "Madding Crowd" to mass gatherings-religion, sport, culture and public health. *Travel Medicine and Infectious Disease* 28, 91–97. DOI: 10.1016/j.tmaid.2018.06.001.

American Telemedicine Association (2020) Telehealth Awareness Week. Available at: https://www.americantelemed.org/ (accessed 15 June 2021).

Bansal, K. (2021) Haridwar Kumbh Mela 2021: Dates, COVID Protocol, Guidelines & Shahi Snan Order. *India.com*, 10 March. Available at: https://www.india.com/news/india/haridwar-kumbh-mela-2021-dates-covid-protocolguidelines-shahi-snan-order-all-details-here-4481213/ (accessed 10 March 2021).

Bhalla, R., Tiwari, P. and Chowdhary, N. (2021) Digital natives leading the world: paragons and values of Generation Z. In: Stylos, N., Rahimi, R., Okumus, B. and Williams, S. (eds) *Generation Z Marketing and Management in Tourism and Hospitality*. Palgrave Macmillan, Cham, Switzerland, pp. 3–23.

Chang, C.L., McAleer, M. and Ramos, V. (2020) A charter for sustainable tourism after COVID-19. *Sustainability (Switzerland)* 12(9), 10–13. DOI: 10.3390/su12093671.

Clara-Rahola, J. (2020) An empirical model for the spread and reduction of the Covid19 pandemic. *Estudios de Economia Aplicada* 38(2). DOI: 10.25115/eea.v38i2.3323.

Collins-Kreiner, N. (2020) Religion and tourism: a diverse and fragmented field in need of a holistic agenda. *Annals of Tourism Research* 82, 102892.

Dwivedi, S. and Cariappa, M.P. (2015) Mass-gathering events: the public health challenge of the Kumbh Mela 2013. *Prehospital and Disaster Medicine* 30(6), 621–624.

Ebrahim, S.H., Ahmed, Q.A., Gozzer, E., Schlagenhauf, P. and Memish, Z.A. (2020) Covid-19 and community mitigation strategies in a pandemic. *BMJ (Clinical Research Ed.)* 368(March), 1–2. DOI: 10.1136/bmj.m1066.

Ebrahim, S.H., Ahmed, Y., Alqahtani, S.A. and Memish, Z.A. (2021) The Hajj pilgrimage during the COVID-19 pandemic in 2020: event hosting without the mass gathering. *Journal of Travel Medicine* 28(2), 1–3. DOI: 10.1093/jtm/taaa194.

Elachola, H., Assiri, A.M. and Memish, Z.A. (2014) Mass gathering-related mask use during 2009 pandemic influenza A (H1N1) and Middle East respiratory syndrome coronavirus. *International Journal of Infectious Diseases* 20(1), 77–78. DOI: 10.1016/j.ijid.2013.12.001.

Gayathri, H., Aparna, P.M. and Verma, A. (2017) A review of studies on understanding crowd dynamics in the context of crowd safety in mass religious gatherings. *International Journal of Disaster Risk Reduction* 25, 82–91. DOI: 10.1016/j.ijdrr.2017.07.017.

Goodwin, H. (2017) The challenge of overtourism. *Responsible Tourism Partnership Working Paper 4,* October.

Griffin, K. and Raj, R. (2017) The importance of religious tourism and pilgrimage: reflecting on definitions, motives and data. *International Journal of Religious Tourism and Pilgrimage* 5(3), 2–9.

Iliev, D. (2020) The evolution of religious tourism: concept, segmentation and development of new identities. *Journal of Hospitality and Tourism Management* 45, 131–140. DOI: 10.1016/j.jhtm.2020.07.012.

Karthika, P.S., Aparna, P.M. and Verma, A. (2018) Understanding crowd dynamics at ghat regions during world's largest mass religious gathering, Kumbh Mela. *International Journal of Disaster Risk Reduction* 31, 918–925. DOI: 10.1016/j.ijdrr.2018.08.005.

Khan, K., Albattah, W., Khan, R.U., Qamar, A.M. and Nayab, D. (2020) Advances and trends in real time visual crowd analysis. *Sensors (Switzerland)* 20(18), 1–28. DOI: 10.3390/s20185073.

Khare, V. (2021) India's Kumbh festival attracts big crowds amid devastating second Covid wave. *BBC.com*, 17 April. Available at: https://www.bbc.co.uk/news/world-asia-india-56770460 (accessed 15 June 2021).

Lee, H. and Graefe, A.R. (2003) Crowding at an arts festival: extending crowding models to the frontcountry. *Tourism Management* 24(1), 1–11. DOI: 10.1016/S0261-5177(02)00036-5.

Manral, K. (2020) "Final Decision on Mahakumbh Mela 2021 to be Taken in February", Says Uttarakhand Chief Minister. *India.com*, 22 June. Available at: https://www.india.com/news/india/final-decisio

n-on-mahakumbh-mela-2021-to-be-taken-infebruary-says-uttarakhand-chief-minister-4064527/ (accessed 16 June 2021).

Mowen, A.J., Vogelsong, H.G. and Graefe, A.R. (2003) Perceived crowding and its relationship to crowd management practices at park and recreation events. *Event Management* 8, 63–72. DOI: 10.3727/152599503108751711.

Neuts, B. and Nijkamp, P. (2012) Crowding perception in a tourist city: a question of preference. *Tinbergen Institute Discussion Paper* 11–140/3. DOI: 10.2139/ssrn.1938174.

Pal, S. (2018) Nashik Teen's Life-Saving Innovation Can Prevent Human Stampedes. *The Better India*, 3 March. Available at: https://www.thebetterindia.com/135438/nilaykulkarni-nashik-innovation-stampede/ (accessed 12 November 2019).

Quadri, S.A. and Padala, P.R. (2021) An aspect of Kumbh Mela massive gathering and COVID-19. *Current Tropical Medicine Reports* 19, 225–230. DOI: 10.1007/s40475-021-00238-1.

Saha, K., Kalra, R. and Khare, R. (2020) A geospatial approach to enhance religious tourism in India—a case of Ujjain city, Madhya Pradesh. *GeoJournal* 1–18. DOI: 10.1007/s10708-020-10334-w.

Schmidt, D. and Keating, J. (1979) Human crowding and personal control: an integration of the research. *Psychological Bulletin* 86(4), 680–700. DOI: 10.1037/0033-2909.86.4.680.

Shinde, K. (2012) Policy, planning, and management for religious tourism in Indian pilgrimage sites. *Journal of Policy Research in Tourism, Leisure and Events* 4(3), 277–301. DOI: 10.1080/19407963.2012.726107.

Singh, S. (2002) Managing the impacts of tourist and pilgrim mobility in the Indian Himalayas / Nouvelles pratiques de tourisme et de pélerinage dans l'Himalaya indien: impacts et modes de gestion. *Revue de Géographie Alpine* 90(1), 25–36. DOI: 10.3406/rga.2002.3070.

Stokols, D. (1972) A social-psychological model of human crowding phenomena. *Journal of the American Planning Association* 38(2), 72–83. DOI: 10.1080/01944367208977409.

Sun, Y.Y. and Budruk, M. (2015) The moderating effect of nationality on crowding perception, its antecedents, and coping behaviours: a study of an urban heritage site in Taiwan. *Current Issues in Tourism* 20(12), 1246–1264. DOI: 10.1080/13683500.2015.1089845.

Tabish, S.A. (2021) India's COVID-19 crisis: challenges and strategies. *International Journal of General Medicine and Pharmacy* 10(1), 86–96.

Tiwari, P., Kainthola, S. and Chowdhary, N.R. (2020) Demarketing: A marketing framework for overtourism. In: Ribeiro de Almeida, C., Quintano, A., Simancas, M., Huete, R., and Breda, Z. (eds) *Handbook of Research on the Impacts, Challenges, and Policy Responses to Overtourism*. IGI Global, London, pp. 94–114. DOI: 10.4018/978-1-7998-2224-0.ch006.

Tomljenović, R. and Dukić, L. (2017) Religious tourism – from a tourism product to an agent of societal transformation. Invited paper presented at: *SITCON 2017: Religious Tourism and the Contemporary Tourism Market*. Singidunum University, Belgrade, pp. 1–8.

UNESCO (United Nations Educational, Scientific and Cultural Organization) (2017) Kumbh Mela – Inscribed in 2017 on the Representative List of the Intangible Cultural Heritage of Humanity. *UNESCO Intangible cultural heritage*. Available at: https://ich.unesco.org/en/RL/kumbh-mela-01258 (accessed 16 June 2021).

Verma, M. and Sarangi, P. (2019) Modeling attributes of religious tourism: a study of Kumbh Mela, India. *Journal of Convention and Event Tourism* 20(4), 296–324. DOI: 10.1080/15470148.2019.1652124.

Vučetić, A. (2021) Analysing religious tourism's influence on tourism policy makers. *International Journal of Religious Tourism and Pilgrimage* 8(8), 1–11.

Yamin, M. (2019) Managing crowds with technology: cases of Hajj and Kumbh Mela. *International Journal of Information Technology (Singapore)* 11(2), 229–237. DOI: 10.1007/s41870-018-0266-1.

Index

Note: Page numbers in *italics* denote tables and figures